Fundamentals of Polar, LDPC and Turbo Codes

Janak Sodha

Fundamentals of Polar, LDPC and Turbo Codes

Including MATLAB, Mathcad and Python Simulations with Desktop Video Lectures

Springer

Janak Sodha
Department of Computer Science,
Mathematics and Physics
The University of the West Indies
Bridgetown, Barbados, WI

ISBN 978-3-031-90923-8 ISBN 978-3-031-90924-5 (eBook)
https://doi.org/10.1007/978-3-031-90924-5

This Springer imprint is published by the registered company Springer Nature Switzerland AG
The registered company address is: Gewerbestrasse 11, 6330 Cham, Switzerland

Preface

This textbook provides a comprehensive introduction to the fundamentals of Polar, LDPC, and Turbo Codes. These codes are essential cornerstones of advanced communication systems. By simplifying the often-intimidating mathematics behind them, this textbook offers clear, step-by-step explanations that assume only a college-level mathematics background.

True understanding comes from engaging with a concept through multiple complementary approaches. To create such an active learning environment, this book offers three integrated pathways. First, full solutions are provided for the problems at the end of each chapter, encouraging students to attempt the questions independently without resorting to AI-generated answers. It is through the act of working through problems on paper that genuine understanding begins to take root. Second, each section is accompanied by a desktop video in which the author covers the core concepts, offering a guided instructor-led experience. Third, simulations in Python, MATLAB, and Mathcad are embedded throughout the chapters to verify and explore the underlying theory. Every detail is inherently visible in a simulation, allowing readers to connect abstract ideas to concrete behavior and to experiment with variations that deepen understanding. Use the QR code below to download the simulations and videos.

https://go.sn.pub/au0ste

Chapter 1 introduces the structure of a digital communication system, focusing on binary phase-shift keying (BPSK) modulation and the additive white Gaussian noise (AWGN) channel model, which form the basis for the analyses in subsequent chapters. This chapter also introduces the concept of error-control coding while maintaining information throughput. Building on the fundamentals, core concepts from information theory, including self-information, entropy, and channel capacity, are introduced. Shannon's channel coding theorem is presented, demonstrating its critical implications for achieving near-error-free communication. Through simulations, the chapter illustrates the performance of BPSK and repetition codes, emphasizing the trade-offs between error correction, signal energy, and information throughput. The role of soft-decision decoding in preserving more capacity is also explored, offering a preview of the advanced coding techniques addressed later in the book.

Chapter 2 explores Polar codes, introduced by Prof. Erdal Arikan in 2009, that have revolutionized error-control codes with their simplicity in encoding and decoding, coupled with capacity-achieving performance. This chapter delves into the fundamental structure of polar codes, explaining them through simple node operations. Detailed demonstrations of encoding and decoding processes for (4, 2) and (8, 4) polar codes are provided. The decoding algorithms are illustrated with examples and simulations, showcasing their error-correction capabilities and the inherent polarization effect, which offers unequal error protection across bits. Beyond the basics, the chapter explores advanced topics such as the reliability sequence derived from the Bhattacharyya parameter and the optimization of belief functions for decoding. The chapter concludes with the inclusion of the 5G standard reliability sequence, providing a comprehensive introduction to polar codes that seamlessly bridges theoretical concepts with practical applications.

Chapter 3 explores Low-Density Parity-Check (LDPC) codes, a class of linear block codes originally developed by Robert Gallager in 1960 and later rediscovered with the emergence of Turbo codes and iterative decoding techniques. LDPC codes, now integral to standards such as IEEE 802.16, IEEE 802.20, and DVB-S2, are widely recognized for their capacity-approaching performance, surpassing Turbo codes in many applications. The chapter begins by introducing the fundamentals of linear block codes, emphasizing the importance of the parity-check matrix and its representation as a Tanner graph. Key decoding algorithms are examined, with a focus on the iterative log-likelihood decoding approach, which balances performance with computational complexity. A detailed case study of a simple regular LDPC code highlights the encoding and decoding processes, supported by simulations. Finally, the design and simulation of WiMAX 802.16e LDPC codes illustrate their practical implementation and performance over noisy channels.

Chapter 4 explores Turbo codes, a very significant advancement in error-control coding that reignited interest in Low-Density Parity-Check (LDPC) codes from the 1960s because of their near-Shannon-limit error-correcting performance. The chapter begins with foundational concepts of convolutional encoding and Viterbi decoding before introducing Recursive Systematic Convolutional (RSC) codes and the Maximum-A-Posteriori (MAP) decoding algorithm. The encoding and decoding processes of Turbo codes are thoroughly explained, highlighting their reliance on iterative decoding using interleaved RSC encoders and MAP decoders. A step-by-step example illustrates the Turbo decoding process, supported by simulation results that showcase the decoder's performance. This in-depth exploration offers a solid grasp of Turbo codes and their implementation.

Finally, I would like to express my heartfelt thanks to my wife Renuka and daughter Sita for their unwavering love, support and patience throughout the writing of this book.

Bridgetown, Barbados, WI Janak Sodha
UWI, 2026

Contents

About the Author

Janak Sodha has over 25 years of experience in telecommunications research and academia. He earned his Ph.D. from the University of Manchester, England. Over his career, he has authored more than 50 publications—spanning both journal articles and conference papers—on topics including convolutional codes, turbo codes, LDPC codes, and polar codes. His work on synchronization techniques has been recognized with citations in several patents.

In 2015, he received the Principal's Award for Excellence for his outstanding contributions to the University and the broader academic community. In 2021, he was elevated to Senior Member of the IEEE, a distinction awarded for his professional standing and significant contributions to electrical and electronic engineering. His leadership at the University of the West Indies includes serving as Head of the Department of Computer Science, Mathematics and Physics (2016–2020).

ORCID: https://orcid.org/0000-0002-9191-4443

List of Figures

List of Simulations

Chapter 1
Introduction

In the early days of communication, systems like the telegraph relied on simple redundancy, repeating symbols or entire messages to detect and correct errors caused by noise and interference. By the 1940s, block codes like Hamming and Golay were developed, using parity checks. In the 1950s, cyclic codes emerged, leading to BCH and Reed–Solomon codes. At the same time, convolutional codes with Viterbi and Fano decoding provided better performance. Block codes continued to evolve, with applications in burst-error correction (e.g., Reed–Solomon) and feedback error control (e.g., CRC codes). In the 1970s and 1980s, concatenated coding schemes were introduced, combining convolutional and block codes for deep space communications. Coded modulation, including techniques like trellis-coded modulation (TCM), also emerged, improving performance by matching coding schemes to modulation formats.

It wasn't until 1993 that error-control codes capable of achieving near-optimal performance emerged, starting with turbo codes, which introduced iterative decoding and led to the rediscovery of LDPC codes from the 1960s. Now we have polar codes, which use the simplicity of repetition codes within a tree structure and information theory principles to achieve near-error-free communication. These modern, powerful codes are the subject of this textbook.

We begin in this chapter with the fundamentals of a digital communication system and present the modulation scheme and channel model assumed throughout the chapters to follow. Then, the concept of error-control coding is introduced, emphasizing how to ensure that the information throughput is not reduced. Finally, the fundamentals of information theory are presented and used to predict the minimum channel signal-to-noise ratio required for near-error-free communication for our assumed modulation scheme and channel model.

J. Sodha, *Fundamentals of Polar, LDPC and Turbo Codes*,
https://doi.org/10.1007/978-3-031-90924-5_1

1.1 Fundamentals

The fundamental blocks of a simple *digital communication system* (DCS) are as follows. The output from an information source (voice, video, data) is represented by binary digits. These binary digits are transferred over a channel by mapping them onto carriers (electromagnetic waves). There are various modulation schemes that could be used to modulate the amplitude, frequency, phase, or a combination of these properties to represent either a single binary digit or strings of binary digits. Given that our focus in this textbook is error-control coding, we shall assume the use of the standard modulation scheme, *Binary Phase-Shift Keying* (BPSK), in which only the phase is used to represent a binary digit. Furthermore, the BPSK signals are transmitted over an *Additive White Gaussian Noise* (AWGN) channel. At the receiver, the coherent demodulator recovers the binary digits by processing the received carrier. These binary digits are then converted back to their original format (voice, video, data) to complete the transfer over the channel. BPSK is covered in detail in Sect. 1.2. Some of the key inherent advantages to representing the information source in a binary format are as follows.

- **Improved Noise Immunity**: Carriers representing binary digits are less susceptible to noise and interference because the demodulator only needs to identify whether a binary zero or one was transmitted, not reproduce the waveform.
- **Combining Information Sources**: Many information sources can be combined and the composite binary stream sent through a given channel.
- **Encryption**: Encryption for secure communication is easily achieved.
- **Efficient Storage**: Binary data can be efficiently stored.
- **Cost-Effectiveness**: Digital components and systems are more cost-effective due to mass production and advancements in semiconductor technology.
- **Error Control Coding**: Extra binary digits can be added to protect against channel errors by using a channel encoder and a corresponding decoder. This is the focus of this textbook.

Without the use of error-control coding, Fig. 1.1 illustrates a simple model of our digital communication system. The binary source represents all the processing blocks (e.g., analog source + source encoder) necessary to transmit information via binary digits x_1 and x_2. The line connecting x_1 to y_1 represents the transmission of the BPSK signal mapped to binary digit 0, which is correctly demodulated at the receiver to output binary digit 0. This line is labeled with the *forward transition probability* $P(0|0)$, which is the probability of demodulating the received BPSK carrier to a binary digit 0, given that a binary digit 0 was transmitted. Similarly, the forward transition probability $P(1|0)$ represents the probability of receiving a binary digit 1, given that a binary digit 0 was transmitted. If $P(1|0) = P(0|1) = \alpha$, the channel is referred to as a *Binary Symmetric Channel* (BSC) and α is known as the BSC *crossover probability*.

In practice, most communication systems can be simplified to the transfer of binary digits over a BSC. In general, for a *discrete memoryless channel* (DMC)

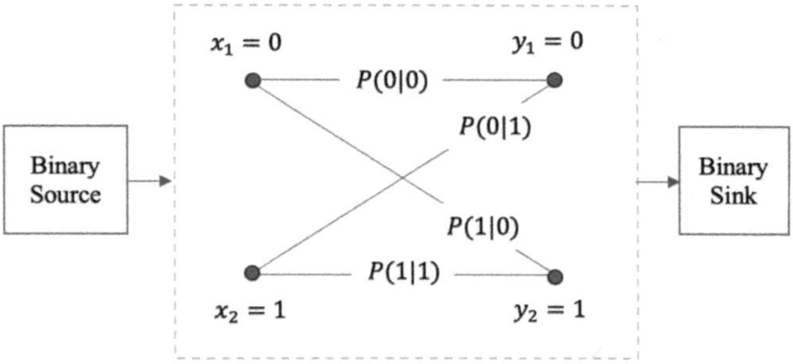

Fig. 1.1 Binary symmetric channel (BSC)

with M possible input symbols and Q possible output symbols, let $P\left(y_j|x_i\right)$ represent the forward transition probability of receiving symbol y_j given that x_i was transmitted, where $i = 1, 2, \ldots, M$ and $j = 1, 2, \ldots, Q$. Also, let $P(x_i)$ represent the probability of transmitting the symbol x_i.

For a BSC, which is a special case of a DMC, $M = Q = 2$ and $P(y_1|x_2) = P(y_2|x_1) = \alpha$. Using these forward transition probabilities and simple probability theory, we can derive some important properties. For instance, in the BSC model, there are only two possible ways for a binary digit to be output in error by the demodulator. Namely, a transmitted binary digit 0 is demodulated as 1 and vice versa. The *probability of an error in a binary digit* P_e at the output of the BSC is therefore given by

$$P_e = P(y_2|x_1)\,P(x_1) + P(y_1|x_2)\,P(x_2) \tag{1.1}$$

where the probability of the joint event that symbol x_1 is transmitted and received as symbol y_2 is $P(y_2|x_1)\,P(x_1)$. Similarly, $P(y_1|x_2)\,P(x_2)$ is the joint event of symbol x_2 transmitted and received as y_1. The plus (+) takes account of either possibility (OR). Given that $P(y_1|x_2) = P(y_2|x_1) = \alpha$ and $P(x_1) + P(x_2) = 1$ (because the transmitted symbol must be either x_1 or x_2), this simplifies to

$$P_e = \alpha P(x_1) + \alpha P(x_2) = \alpha\left[P(x_1) + P(x_2)\right] = \alpha \tag{1.2}$$

Notice that P_e does not depend on $P(x_1)$ and $P(x_2)$ if the channel is symmetric, i.e., $P(y_1|x_2) = P(y_2|x_1)$. Finally, given the fact that $P(y_1|x_2) + P(y_2|x_2) = 1$ and $P(y_2|x_1) + P(y_1|x_1) = 1$, we have for a BSC, $P(y_1|x_1) = P(y_2|x_2) = 1 - \alpha$.

It is important to understand that, for example, $P_e = 0.1$ does **not** imply that an error will occur exactly once every 10 binary digits. When observing the decoded information sequence in blocks of 10 binary digits, we might find that some blocks contain no errors, while others may have three or four errors randomly distributed. However, when averaged over a large number of blocks, e.g., 2000 blocks (equivalent to 20,000 binary digits), the ratio of the total number of binary digit errors to the total number of transmitted binary digits will converge to approximately 0.1. This ratio is referred to as the *Bit Error Rate* (BER). As more blocks are processed, the BER estimation becomes more accurate, as will be demonstrated next in the simulation of a Binary Symmetric Channel (BSC).

1.1.1 Simulation: BSC

In simulation BSC, the digital communication system (DCS) shown in Fig. 1.1 is simulated. A uniform *random number generator* (RNG) is used to generate a binary digit zero with probability p as illustrated in Fig. 1.2. The rectangular box represents the range of values that the RNG can produce, uniformly distributed between 0 and 1. A threshold p is set within this range. The number r generated by the RNG will lie between 0 and p with probability p and the remaining interval with probability $(1 - p)$. Thus, a binary digit 0 is output if $r \leq p$; otherwise, a binary digit 1 is output. To implement the BSC, another RNG is used with a threshold set to α. If $r \leq \alpha$, the input binary digit is inverted; otherwise, it is allowed to pass through unchanged. Simulation results are presented in Fig. 1.3. As expected from Eq. 1.2, the probability of an error in a binary digit at the sink, $P_e = \alpha$.

Experiment Alter the value of p and confirm for yourself that it makes no difference to the estimate of P_e for a given value of α.

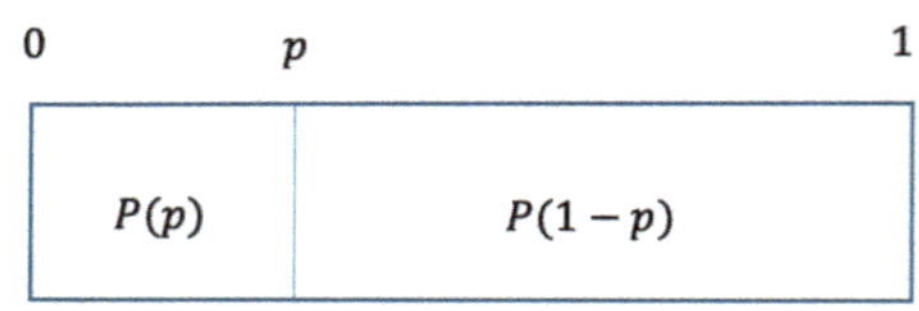

Fig. 1.2 Uniform random number generator

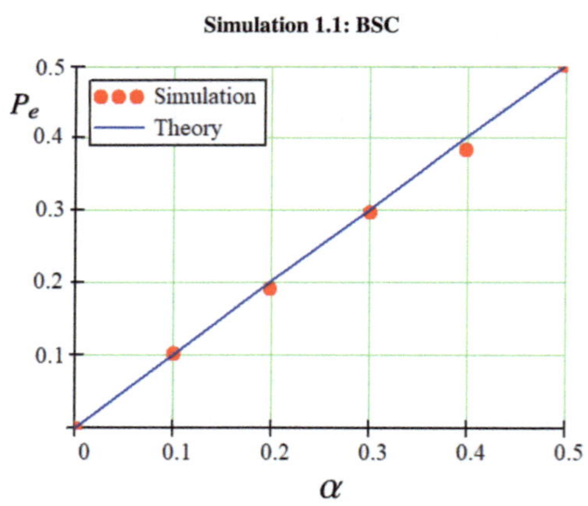

Fig. 1.3 Binary symmetric channel (BSC)

1.2 Binary Phase-Shift Keying (BPSK)

In this section, the fundamentals of binary phase-shift keying (BPSK) and its performance over an AWGN channel are covered briefly to establish notation and help explain the impact of error-control coding on the overall BER. Suppose the binary digits 0 and 1 are represented by the BPSK signals $s_1(t) = A\cos(2\pi f_c t)$ and $s_2(t) = -A\cos(2\pi f_c t)$, respectively, where A is the amplitude and f_c is the frequency of the sinusoidal carrier signal. It can be easily shown (problem 2) that the energy of a BPSK signal is $E_b = \frac{A^2 T_b}{2}$, where T_b is the time duration of the signal. Expressing these signals in terms of the *basis function* $\phi_1(t)$ defined by

$$\phi_1(t) = \sqrt{\frac{2}{T_b}}\cos(2\pi f_c t) \tag{1.3}$$

we find $s_1(t) = \phi_1(t)\sqrt{E_b}$ and $s_2(t) = -\phi_1(t)\sqrt{E_b}$. At the receiver, a coherent demodulator multiplies the incoming signal $r(t)$ with basis function $\phi_1(t)$ and integrates over the time duration T_b of the signal to output the received *signal point coordinate* x_h. Namely,

$$x_h = \int_0^{T_b} r(t)\phi_1(t)\,dt \tag{1.4}$$

It is easily shown (problem 2) that for a noiseless channel in which $r(t)$ is either $s_1(t)$ or $s_2(t)$, the signal point coordinate x_h is either $\sqrt{E_b}$ or $-\sqrt{E_b}$, respectively. If $x_h \geq 0$, the demodulator will output a binary digit 0, otherwise a binary digit 1. In this case, the demodulator is said to have made a *hard-decision*. To determine an expression for the probability of making an error, we need to consider the *probability density functions* (PDFs), which govern the two signal points as shown in Fig. 1.4a, given by

$$f\left(x|\sqrt{E_b}\right) = \frac{1}{\sigma\sqrt{2\pi}}\exp\left[\frac{-\left(x-\sqrt{E_b}\right)^2}{2\sigma^2}\right] \tag{1.5}$$

$$f\left(x|-\sqrt{E_b}\right) = \frac{1}{\sigma\sqrt{2\pi}}\exp\left[\frac{-\left(x+\sqrt{E_b}\right)^2}{2\sigma^2}\right] \tag{1.6}$$

where for coherent demodulation, it can be shown [1] that

$$\sigma = \sqrt{\frac{N_o}{2}} \tag{1.7}$$

where N_o is the *single-sided noise power spectral density*.

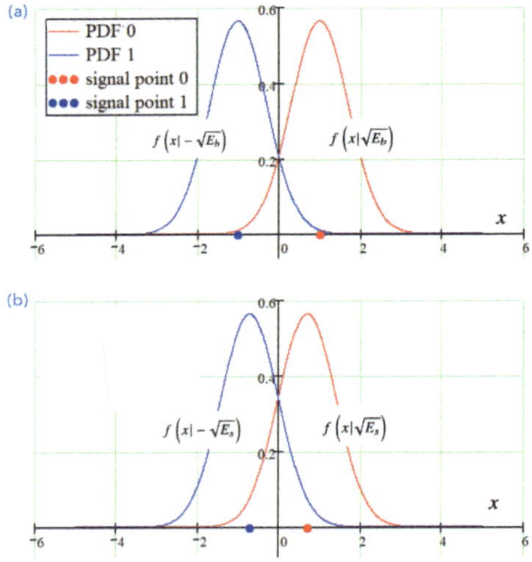

Simulation 1.2: BPSKfundmentals

Fig. 1.4 BPSK Signal Point and PDFs (**a**) Uncoded $E_s = E_b$ (**b**) Coded $E_s = E_b/2$

The PDFs and corresponding signal points in Fig. 1.4b are shown in this figure for ease of comparison when we consider the impact of error-control coding on the information throughput in Sect. 1.4. The area under a given PDF represents probability. For example, the total area under $f\left(x|\sqrt{E_b}\right)$ is the probability that the signal point lies somewhere along the x-axis, which is of course 1 and given by

$$\int_{-\infty}^{+\infty} f\left(x \mid \sqrt{E_b}\right) dx = 1 \tag{1.8}$$

In Fig. 1.4a, $f\left(x| - \sqrt{E_b}\right)$ is centered on the signal point located at $-\sqrt{E_b}$. Notice that a part of this curve extends into the region $x_h \geq 0$. The area under this curve in this region is the probability that the signal point will have a coordinate $x_h \geq 0$. Thus, the forward transition probability $P(0|1)$ is given by

$$P(0|1) = \int_{0}^{+\infty} f\left(x \mid -\sqrt{E_b}\right) dx \tag{1.9}$$

which can be shown (problem 3) to be given by

$$P(0|1) = Q\left(\sqrt{\frac{2E_b}{N_o}}\right) \tag{1.10}$$

where E_b is the energy per BPSK signal and the $Q(.)$ function is defined by

$$Q(u) = \frac{1}{\sqrt{2\pi}} \int_u^{\infty} e^{-\frac{z^2}{2}} dz \tag{1.11}$$

By symmetry, $P(0|1) = P(1|0) = \alpha$. Let SNR represent the ratio E_b/N_o in decibels (dB) so that

$$SNR = 10 \log_{10} \left(\frac{E_b}{N_o} \right) \tag{1.12}$$

from which $\frac{E_b}{N_o} = 10^{\frac{SNR}{10}}$. Therefore, the standard deviation σ expressed in terms of the channel signal-to-noise ratio (SNR) in decibels is

$$\sigma = \sqrt{\frac{E_b}{2 \cdot 10^{\frac{SNR}{10}}}} \tag{1.13}$$

and the corresponding BSC crossover probability α is given by

$$\alpha = Q\left(\sqrt{\frac{2E_b}{N_o}}\right) = Q\left(\sqrt{2 \cdot 10^{\frac{SNR}{10}}}\right) \tag{1.14}$$

The variation of this probability, that is, the performance of *uncoded* BPSK, is shown in Fig. 1.6 over a range of SNRs. This will be verified by simulation in the next section. Note that in Sect. 1.3.2, we shall find that α becomes **worse** with the use of error-control coding at a given SNR.

1.2.1 Simulation: Uncoded BPSK

In simulation BPSK, the digital communication system (DCS) shown in Fig. 1.5 is simulated, where we expect (from Eq. 1.2) the probability of an error in a binary digit at the sink, $P_e = \alpha = Q\left(\sqrt{2 \cdot 10^{\frac{SNR}{10}}}\right)$. To model the AWGN noise on the signal points located at $\pm\sqrt{E_b}$, the uniform random number generator output r is used to create a Gaussian random number generator with a mean m and standard

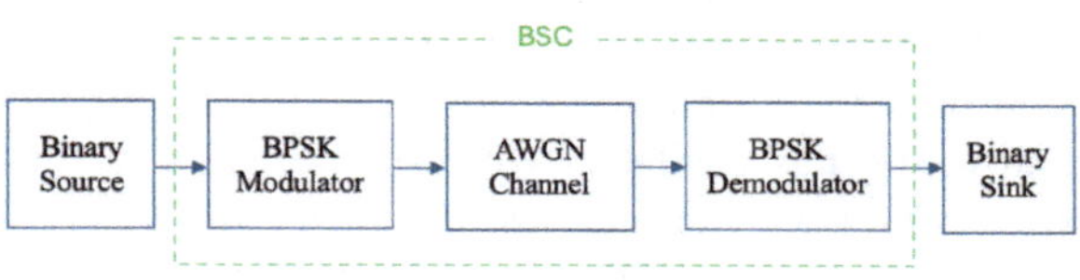

Fig. 1.5 Uncoded BPSK over an AWGN channel

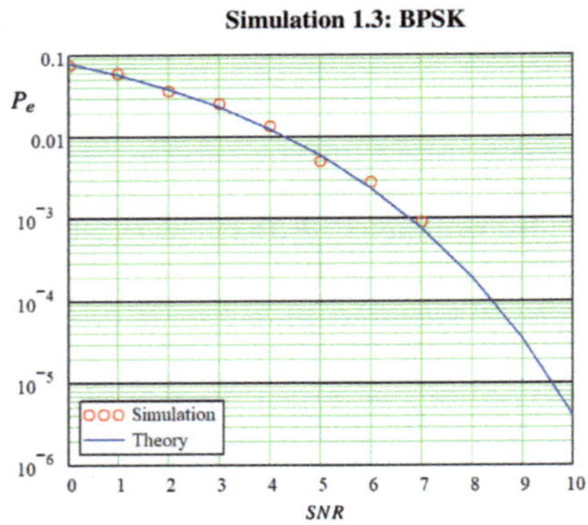

Fig. 1.6 Uncoded BPSK

deviation σ via the formula

$$\text{GAUSS}(m, \sigma) = m + \sigma\sqrt{-2\ln(r)} \cdot \cos(2\pi r) \tag{1.15}$$

where the value of r, used twice in this equation, is not the same value because each time the RNG outputs a random number. The horizontal coordinate x_h of the received signal point is thus given by

$$x_h = GAUSS(\pm\sqrt{E_b}, \sigma) \tag{1.16}$$

where, for example, $\sqrt{E_b}$ corresponds to a binary digit 0 being transmitted. If $x_h \geq 0$, the BPSK demodulator output is a binary digit 0; otherwise, it outputs a binary digit 1. A comparison between the theoretical and simulated performance is shown in Fig. 1.6, where we observe excellent agreement.

1.3 Error-Control Coding

To reduce P_e, we could simply increase the energy E_b of a BPSK signal by increasing its amplitude A or its time duration T_b. However this is not practical. For example, the necessary battery size and weight will be unacceptable for a mobile phone. The solution is to introduce *error-control coding* (also referred to as *channel coding*) as shown in Fig. 1.7, in which a channel encoder and its corresponding decoder are inserted on either side of the channel.

Error-control coding could be used to only *detect* errors within a given received block. If an error is detected, then an *automatic repeat request* (ARQ) signal is sent back to the transmitter to retransmit if a duplex link between the transmitter and receiver is available. However, we shall restrict ourselves to *forward error-correction* (FEC) techniques, in which extra binary digits are added by a channel encoder prior to transmission, to create a code structure that is exploited by a channel

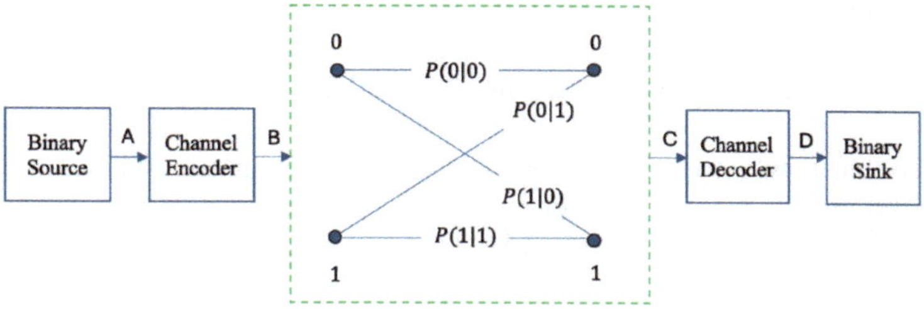

Fig. 1.7 Channel coding over a BSC

decoder to correct as many channel errors as possible and output the estimated information binary sequence.

The aim is to reduce P_e at the binary sink (point D) without affecting the *information throughput*. To understand this concept, suppose the channel encoder in Fig. 1.7 uses a simple *repetition* code, which is the simplest FEC error-control code. Its encoding algorithm is as follows: if the binary source outputs a 1, the channel encoder outputs N binary digit 1s; if the source outputs a 0, the encoder outputs N binary digit 0s. Let m represent the input binary digit and c represent the N-digit *codeword* output from the channel encoder. For example, if $N = 3$ and $m = 0$, then $c = [000]$ and if $m = 1$, $c = [111]$. The individual binary digits within the codeword c are referred to as *code digits*. In general, let K represent the number of binary digits input to the channel encoder, which then outputs N code digits. The *code rate* R_{code} is defined to be the ratio $\frac{K}{N}$. Thus, for a repetition code, the code rate $R_{code} = \frac{1}{N}$.

If N is an odd number with $N \geq 3$, the channel decoder can use majority logic decoding as follows. Count the number of binary digit 1s and 0s within the received word. If the number of 1s is greater than the number of 0s, then output a binary digit 1; otherwise, output a binary digit 0. Table 1.1 illustrates the encoding and decoding processes by example for $N = 3$ for the input information binary stream 1011.

At point A within Fig. 1.7, the information binary digits 1011 are output from the binary source. These are then encoded by the channel encoder to output the codewords at point B. These code digits are sent over the BSC one by one. The binary sequence at point C is an example of what could be received at the output of

Table 1.1 FEC example using a rate 1/3 repetition code over a BSC

Point					Description
A	1	0	1	1	Information binary digits
B	111	000	111	111	Code digits transmitted
C	110	110	011	101	Received binary digits from BSC channel
D	1	1	1	1	Decoded information binary digits

the BSC. Notice that five code digits were received in error from Table 1.1. However, at point D after decoding using majority logic decoding, only one error occurred in the received information binary stream, i.e., we have achieved FEC error-control coding.

Notice that if two code digits are corrupted within a given codeword, then the decoder will make an error. If the entire codeword is received in error, for example, [000] is transmitted and received as [111], the decoder will make an error even though a valid codeword has been received. This is an unavoidable source of error in all error-control codes.

If $N = 1$, the probability of an error in a binary digit P_e at the sink will be equal to the BSC crossover probability. As N is increased, P_e will be reduced as will be demonstrated by simulation in Sect. 1.3.3. So why not simply increase N from 3 to 51? Although the error-control capability rapidly improves as N is increased, such an error control code is not used in practice because the information throughput should **not** be reduced by the use of a channel encoder and decoder combination. Namely, the TV news broadcast at 9 p.m. should not be received late by several minutes, but under 30 ms to avoid the human brain detecting a time delay. To clarify this further, suppose it takes 1 second for a binary digit to be transferred through the BSC. Without the use of error-control coding, it would take 10 seconds to transfer 10 information binary digits across the BSC. If a rate 1/3 repetition error-control code is used without any changes, it would take 30 seconds to transfer the 30 code digits across the BSC because the code digits are transferred over the BSC, one by one. For the information throughput to be unaffected, the 30 code digits must be transferred in 10 seconds. We shall take a closer look at this very important constraint in error-control coding next.

1.3.1 Information Throughput

Consider the DCS shown in Fig. 1.8 in which r_b binary digits per second are input to the channel encoder, which in turn outputs r_c code digits per second.

To ensure that the information throughput remains constant, the time-occupancy $T_b K$ occupied by the K information digits fed into the encoder must be equal to the time-occupancy $T_c N$ occupied by the N code digits output, where $T_b = \frac{1}{r_b}$ and $T_c = \frac{1}{r_c}$. This relationship is expressed as

$$T_b K = T_c N \tag{1.17}$$

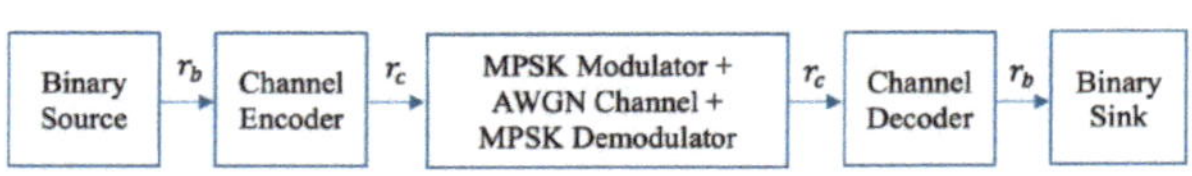

Fig. 1.8 Coded MPSK over an AWGN channel

from which

$$T_c = R_{code} T_b \tag{1.18}$$

because $R_{\text{code}} = \frac{K}{N}$. In general, using *M-ary Phase-Shift Keying* (MPSK), there are M possible signals, each with a carrier phase set to one of M distinct values. A total of $\log_2 M$ code digits can be mapped onto an MPSK signal. For example, BPSK is simply MPSK with $M = 2$. The time-occupancy $T_c N$ occupied by the N code digits is equal to the time duration T_s of an MPSK signal, and each MPSK signal maps to $\log_2 M$ code digits. Thus,

$$T_s = T_c N = R_{code} T_b N = R_{code} T_b \log_2 M \tag{1.19}$$

For example, in *Quadrature Phase-Shift Keying* (QPSK), $M = 4$ and there are four QPSK signals with a carrier phase set to one of four possible values (e.g. 0, $\pi/2$, π, $3\pi/2$). Each QPSK signal is mapped to $N = \log_2 M = \log_2 4 = 2$ code digits, meaning the QPSK signals are mapped to the binary pairs 00, 01, 10, and 11. The time duration of a QPSK signal is $T_s = R_{code} T_b \log_2 M = 2T_b R_{code}$. If the channel encoder is removed, then $R_{code} = 1$ and $T_s = 2T_b$. Thus, a QPSK signal now has twice the time duration of a BPSK signal and represents two binary digits. If the amplitude A and frequency f_c are the same as those of a BPSK signal, the QPSK signal will have twice the energy of a BPSK signal (problem 4). In general, the energy of an MPSK signal (problem 7) is given by

$$E_s = R_{code} E_b \log_2 M \tag{1.20}$$

For BPSK, $M = 2$ and $T_s = R_{code} T_b$ to ensure that the information throughput is unaffected, i.e., the time duration of the BPSK is simply reduced by a factor of R_{code}. Note that a BPSK signal now represents a code digit. Following the same analysis as before, the BSC crossover probability using BPSK over an AWGN channel is now given by

$$\alpha = Q\left(\sqrt{\frac{2E_s}{N_o}}\right) \tag{1.21}$$

where $E_s = R_{code} E_b$ is the energy of a BPSK signal. To summarize, if the channel encoder and decoder are removed, so that $R_{code} = 1$, then a BPSK signal represents an information binary digit with $E_s = E_b$. Otherwise, with $R_{code} < 1$, a BPSK signal represents a code digit, and the energy E_s of a BPSK signal is reduced by a factor of R_{code}. To understand the consequences of this, the signal constellation diagram of BPSK is presented in Fig. 1.4b, overlaid with the corresponding probability density functions assuming an AWGN channel. The locations of the two signal points are now $\pm\sqrt{E_s}$. For $R_{code} = 1$, they are located at $\pm\sqrt{E_b}$ as shown in Fig. 1.4a. However, for $R_{code} = 1/2$, the locations change to

$\pm\sqrt{E_b/2}$ and the two signal points move closer together, thereby increasing the BSC crossover probability as evident in Fig. 1.4b by the larger area of overlap. A plot of Eq. 1.21 is presented in Fig. 1.9 for $R_{code} = 1$ (uncoded BPSK) and $R_{code} = 1/2$ for a range of SNRs. Note the increase in the BSC crossover probability α for $R_{code} = 1/2$, making code digits more susceptible to noise. This is as expected because each BPSK signal now has less energy. Even though more code digits may be received in error, the channel decoder will use the redundancy introduced by the channel encoder to reduce P_e at the binary sink below what it would have been without the use of error-control coding. As a standard rule of thumb in digital communication systems, we aim to ensure that the probability of error P_e at the binary sink is reduced below 10^{-6}. This corresponds to an average of one error per 1 million information bits processed through the system. For the rate 1/3 repetition code example, switching to 8PSK to send the three code digits makes it worse due to the reduced distance between signal constellation points (problem 8).

1.3.2 Simulation: Coded BPSK

In simulation-Coded BPSK, the simulation is similar to that in Sect. 1.2.1. The corresponding simulation results overlaid on the theoretical curve are presented in Fig. 1.9. The difference is that $\alpha = Q\left(\sqrt{\frac{2E_s}{N_o}}\right) = Q\left(\sqrt{\frac{2R_{\text{code}}E_b}{N_o}}\right)$. To ensure that the signal points now located at $\pm\sqrt{E_s}$ remain at ± 1, we may set $E_b = \frac{1}{R_{\text{code}}}$ without any loss of generality. Therefore,

$$\sigma = \sqrt{\frac{E_b}{2 \cdot 10^{\text{SNR}/10}}} = \sqrt{\frac{1}{2R_{\text{code}} \cdot 10^{\text{SNR}/10}}} \tag{1.22}$$

Notice that as R_{code} is reduced, the standard deviation increases, which corresponds to more noise on a signal point.

Experiment Modify the simulation as follows. Move the signal points closer together by setting the location of the signal points to $\pm\sqrt{E_s}$, where $E_s = R_{code}E_b$ and maintain $\sigma = \sqrt{\frac{E_b}{2 \cdot 10^{SNR/10}}}$ as R_{code} is changed. For example, set $E_b = 4$ joules and $R_{code} = 1/2$. You should recover the same simulation results as in Fig. 1.9.

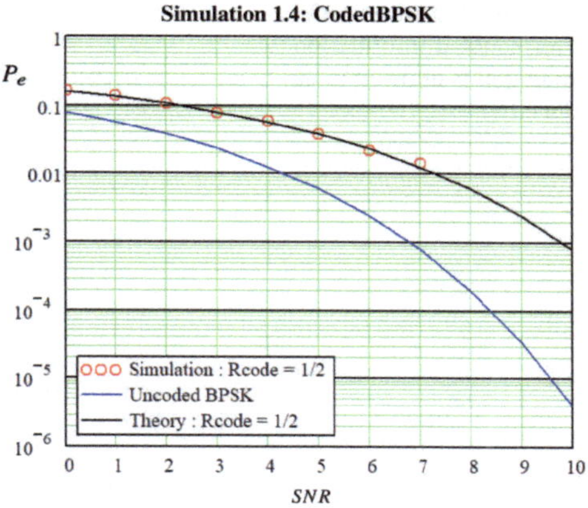

Fig. 1.9 BPSK performance over an AWGN channel with a $R_{code} = 1/2$ channel encoder

1.3.3 Simulation: Repetition Code

In simulation-Repetition Code, the DCS shown in Fig. 1.8 is simulated in which the channel encoder is a rate $1/N$ repetition code. The combination of the MPSK modulator, AWGN channel, and MPSK demodulator is replaced with a BSC. Binary digits are processed until a specified number of errors N_{errors} occur at the binary sink. If the information throughput is **allowed to be reduced**, then it can be shown (problem 10) that P_e at the sink is given by

$$P_e = \sum_{i=\frac{(N-1)}{2}+1}^{N} \frac{N!}{(N-i)!i!} \alpha^i (1-\alpha)^{N-i} \tag{1.23}$$

Using this theoretical expression, Fig. 1.10 shows the correspondence between simulation and theory for two repetition codes of rate 1/3 and 1/9 over a range of α values. On the vertical axis is P_e at the binary sink and on the horizontal axis is the BSC crossover probability α. We have excellent agreement between the simulation and theoretical curves. For $\alpha = 0.1$, using a rate 1/9 repetition code, P_e is reduced to approximately 10^{-3} if the information throughput is reduced.

- **Experiment**: Modify the simulation code to consider what happens if the information throughput is **not** allowed to be reduced.

1.4 Information Theory

In this section, the topic of information theory is presented in brief. Developed by Claude Shannon [2], information theory is the foundation of error-control coding. Specifically, we will determine the minimum channel signal-to-noise ratio required

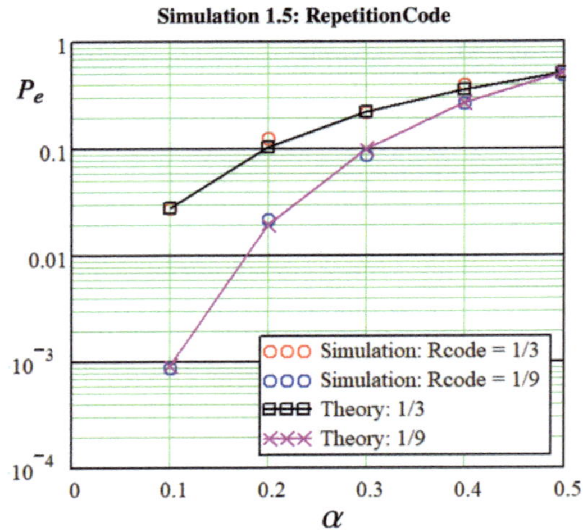

Fig. 1.10 Repetition code performance if information throughput is allowed to be reduced

for near-error-free communication using BPSK over an AWGN channel model. A central result of this theory is Shannon's channel coding theorem, which has motivated and continues to inspire research in the field of error-control coding for over 60 years.

1.4.1 What is Information?

If you are told that the sun will "rise" tomorrow, whether visible or obstructed by clouds, no information has been conveyed because the probability of that event is 1 (thankfully, for the foreseeable future). Conversely, if you are told that an earthquake will occur tomorrow at your home, a large amount of information has been conveyed because the probability of this event is very small. Accordingly, Claude Shannon defined the *self-information* I associated with an event A to be given by

$$I = \log_2 \frac{1}{P(A)} \text{ bits} \tag{1.24}$$

where $P(A)$ is the probability of the event A. The units of information are *bits*, derived from the use of the base-2 logarithm, which relates to binary digits as we will discover in Sect. 1.4.5. Using the $\log_2(.)$ function, the total information conveyed by two independent events A and B is simply the sum of the information from each event, i.e., $\log_2 \frac{1}{P(A)} + \log_2 \frac{1}{P(B)} = \log_2 \frac{1}{P(A)P(B)}$. The variation of I versus $P(A)$ is presented in Fig. 1.11. As required, self-information I is a positive quantity and as $P(A)$ tends toward 1, I tends toward 0.

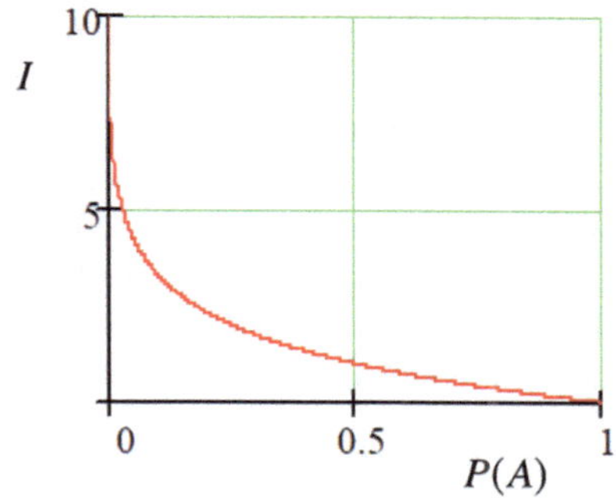

Fig. 1.11 Self-Information $I = \log_2 \frac{1}{P(A)}$ bits

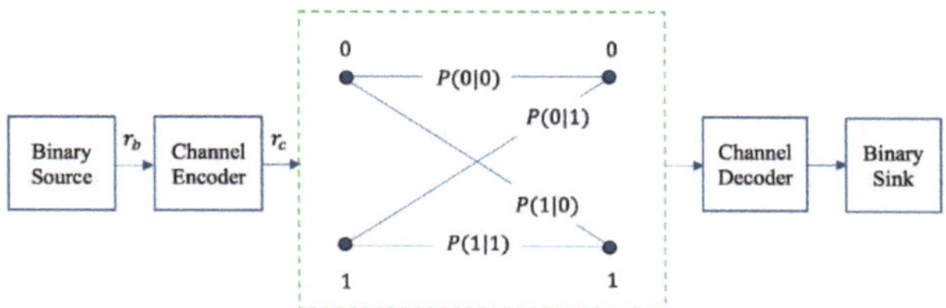

Fig. 1.12 Channel coding over a BSC

1.4.2 Source Entropy and Information Rate

For the binary source in Fig. 1.12, let r_b represent the number of binary digits per second output, and let $P(x_i)$ represent the probability with which x_i is output by the binary source. The information conveyed by a given binary digit is $\log_2 \frac{1}{P(x_i)}$ bits. Taking the probability-weighted average, the average information per binary digit, referred to as the *source entropy* $H(X)$, is given by

$$H(X) = \sum_{i=1}^{M} P(x_i) \log_2 \frac{1}{P(x_i)} \text{bits per symbol} \tag{1.25}$$

where X represents the set of source symbols, i.e., x_1, x_2. A symbol is a binary digit in this case. Using Eq. 1.25 for $P(0) = p$, the source entropy $H(X) = \Omega(p)$, where $\Omega(.)$ is the *binary entropy function* is defined by

$$\Omega(z) = z \log_2 \left(\frac{1}{z}\right) + (1 - z) \log_2 \left(\frac{1}{1 - z}\right) \tag{1.26}$$

The *information rate* R (bits/sec) of this source is defined as

$$R = r_b H(X) = r_b \Omega(p) \text{ bits per second} \tag{1.27}$$

To understand this quantity R, think of it as the flow of water and visualize the channel as a water pipe connected from the source to the sink. If the rate at which water flows out of the source is less than or equal to the capacity of the channel pipe, it should be possible to transfer all the water (information) across the channel with minimal leakage (essentially error-free). However, if the rate at which water flows out of the source exceeds the capacity of the pipe, some water will leak. The information rate R, measured in bits per second, lies at the heart of Shannon's channel coding theorem, which will be presented in Sect. 1.4.5.

1.4.3 Average Mutual Information: DMC

For a discrete memoryless channel (DMC), the *average mutual information* $I(X, Y)$, which represents the average information transferred per symbol, is given by

$$I(X, Y) = \sum_{i=1}^{M} \sum_{j=1}^{Q} P(x_i, y_j) I(x_i, y_j) \text{ bits/symbol}$$

where the *mutual information* $I(x_i, y_j)$ is defined as

$$I(x_i, y_j) = \log_2 \frac{P(y_j|x_i)}{P(y_j)} \text{ bits}$$

Notice the definition of $I(X, Y)$ is analogous to Eq. 1.25. Here, $P(x_i, y_j)$ is the joint probability of input x_i and output y_j, while $P(y_j|x_i)$ is the conditional probability of y_j given x_i.

For a noiseless channel, $y_j = x_i$ and $P(y_j|x_i) = 1$ for $i = j$. In this case, the information transferred across the channel simplifies to $\log_2 \frac{1}{P(x_i)}$, which is the self-information of each input symbol. However, for a very noisy channel, x_i and y_j are independent. In this case, $P(y_j|x_i) = P(y_j)$ and the mutual information becomes

$$I(x_i, y_j) = \log_2 \frac{P(y_j|x_i)}{P(y_j)} = \log_2 \frac{P(y_j)}{P(y_j)} = 0$$

As expected, no information is transferred across the channel. It can be shown [4] that the average mutual information is given by

$$I(X, Y) = H(Y) - H(Y|X) \text{ bits/symbol} \tag{1.28}$$

where $H(Y)$ is the average information per received symbol, given by

$$H(Y) = \sum_{j=1}^{Q} P(y_j) \log_2 \frac{1}{P(y_j)} \tag{1.29}$$

and $H(Y|X)$ is the average information per symbol added by the channel, given by

$$H(Y|X) = \sum_{i=1}^{M} \sum_{j=1}^{Q} P(x_i) P(y_j|x_i) \log_2 \frac{1}{P(y_j|x_i)} \tag{1.30}$$

For a noiseless channel, $H(Y|X) = 0$ and as expected, $I(X, Y) = H(Y) = H(X)$. Finally, the channel capacity C_s in bits per symbol for a DMC, is the maximum value of $I(X, Y)$. We shall apply these formulas in the next section where we consider the channel capacity of a BSC.

1.4.4 Channel Capacity of a BSC

A symbol is a binary digit in the case of a BSC. It is easily shown (problem 11) that the maximum average information transferred across a BSC per symbol C_s is given by

$$C_s = [1 - \Omega(\alpha)] \text{ bits/symbol} \tag{1.31}$$

The variation of this function is shown in Fig. 1.13. For maximum noise where $\alpha = 0.5$, information is not transferred across the channel because $\Omega(0.5) = 1$ and $C_s = 0$ bits/symbol. For a noiseless channel where $\alpha = 0$, $C_s = 1$ bit/symbol as expected because the maximum possible binary source entropy is 1 bit/symbol for $p = 0.5$. For a transmission rate of r_c binary digits per second over a BSC, the

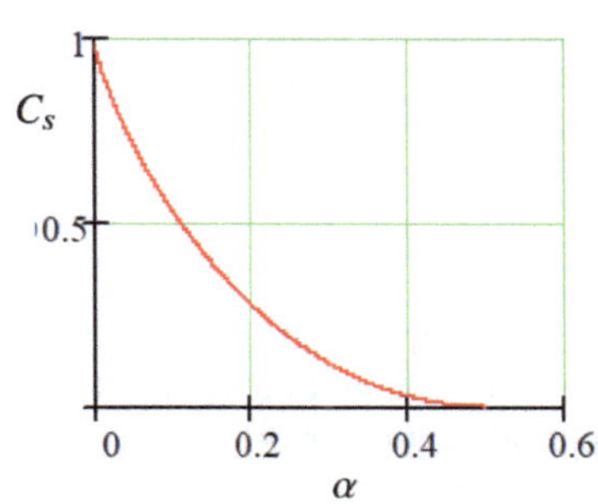

Fig. 1.13 Variation of the channel capacity C_s in bits/symbol versus BSC crossover probability α

channel capacity C, which represents the maximum achievable information transfer rate, is given by

$$C = r_c C_s \text{ bits/sec} \tag{1.32}$$

In Eq. 1.31, the crossover probability α depends on the type of modulator, physical channel, and the demodulator. We shall restrict ourselves to the use of BPSK over an AWGN channel throughout the chapters to follow.

1.4.5 Shannon's Channel Coding Theorem

For reliable communication, Shannon's channel coding theorem requires the source information rate $R \leq C$. Conversely, if $R > C$, it is not possible to transmit information over the channel and reconstruct it with an arbitrarily small probability of error. Note carefully that Shannon's channel coding theorem does **not** provide a method to design the appropriate channel encoder and decoder. For example, consider the binary source in Fig. 1.12 that outputs a binary digit 0 with probability p at a rate of r_b binary digits per second. These are encoded by a channel encoder to produce r_c code digits per second, which are then transmitted over the channel. However, the binary source entropy is $\Omega(p)$, and its information rate is $R = r_b\Omega(p)$ bits per second. Thus, Shannon's channel coding theorem requires

$$r_b\Omega(p) \leq C \tag{1.33}$$

Using source coding, it is not difficult to achieve p near 0.5. For $p = 0.5$, $\Omega(0.5) = 1$ and the theorem requires that $r_b \leq C$. This means that the maximum rate at which binary digits can be transferred over the channel essentially error-free is C binary digits per second. Furthermore, substituting Eq. 1.32 into Eq. 1.33, we get

$$r_b\Omega(p) \leq r_c C_s \tag{1.34}$$

Given that $R_{\text{code}} = \frac{r_b}{r_c}$ and assuming $p = 0.5$, we find the requirement for reliable communication to be

$$R_{\text{code}} \leq C_s \tag{1.35}$$

The implication of this result is significant! Namely, that there exists an error-control code with a code rate $R_{\text{code}} \leq C_s$ capable of achieving an arbitrarily low probability of error, e.g., $P_e < 10^{-18}$. Equivalently, provided $C_s \geq R_{\text{code}}$, near-error-free communication is possible. Using hard-decision BPSK over an AWGN channel, Shannon's channel coding theorem can predict the minimum channel SNR

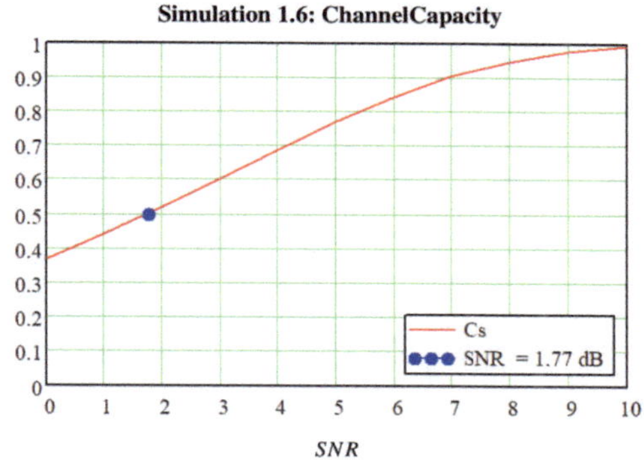

Fig. 1.14 Channel capacity C_s in bits/symbol for BPSK over an AWGN with hard-decision

(dB) required for virtually error-free communication as follows. In this case,

$$C_s = 1 - \Omega\left(Q\left(\sqrt{\frac{2E_b R_{\text{code}}}{N_o}}\right)\right) \tag{1.36}$$

For $R_{\text{code}} = \frac{1}{2}$, Fig. 1.14 shows the variation of C_s with the channel SNR. For $\left(\frac{E_b}{N_o}\right)$ dB $= 1.77$ dB, $R_{\text{code}} = C_s$. More importantly, provided $C_s \geq \frac{1}{2}$, or equivalently the channel SNR is greater than or equal to 1.77 dB, there exists an error-control coding scheme utilizing a rate 1/2 code that can achieve virtually error-free communication over an AWGN channel using BPSK with hard-decision decoding.

1.4.6 Soft-decision

Up to this point, we have restricted ourselves to a BSC in which the output from the BPSK demodulator x_h is hard-decisioned to output either a binary digit 0 or 1. When making a hard decision, valuable information is lost because the demodulator output is forced into one of two binary states. A better approach is to make *soft-decisions* by quantizing the demodulator output into multiple levels. For example, using 8-level soft decision where $M = 2$ and $Q = 8$, integers 0–7 are output by the demodulator as shown in Fig. 1.15.

The integers 0–3 represent a binary digit 0 and 4–7 represent a binary digit 1, with 3 and 4 representing weak binary digits 0 and 1, respectively. It can be shown [3] that in general for Q_{level} of quantization, the channel capacity is given by

$$C_s = \sum_{j=0}^{Q_{\text{level}}-1} P(j|0)\log_2\left(\frac{2P(j|0)}{P(j|0)+P(j|1)}\right) \text{ bits/symbol} \tag{1.37}$$

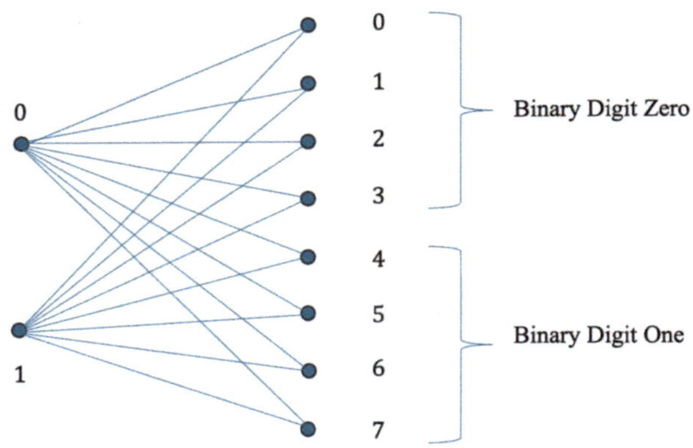

Fig. 1.15 8-level quantization

Taking this one step further, if the output of the demodulator is not quantized (infinite level of quantization), the capacity formula for the binary-input AWGN channel can be shown [4] to be given by

$$C_s = \frac{1}{2}\sum_{k=0}^{1}\int_{-\infty}^{\infty} p(y|k)\log_2\left(\frac{p(y|k)}{p(y)}\right) dy \tag{1.38}$$

where $p(y|k)$ is the conditional PDF

$$p(y|k) = \frac{1}{\sqrt{2\pi\sigma^2}} \exp\left(-\frac{(y-m_k)^2}{2\sigma^2}\right) \tag{1.39}$$

and $p(y) = \frac{1}{2}\left(p(y|0) + p(y|1)\right)$. For BPSK, $m_k = \pm\sqrt{R_{\text{code}}E_b}$ and $\sigma^2 = \frac{N_o}{2}$ in Eq. 1.39. Using the Eqs. 1.36, 1.37 and 1.38, Fig. 1.16 shows a comparison between these curves for the minimum SNR at which $C_s = R_{\text{code}}$.

As R_{code} approaches 0, the minimum SNR required to achieve capacity is -1.59 dB. The curves show that there is little difference between using 8-level quantization and infinite-level quantization at higher SNRs. However, there is much to be gained by switching from hard-decision to soft-decision decoding. For example, the minimum SNR required for essentially error-free communication using a rate 1/2 code is 1.77 dB, 0.31 dB, or 0.19 dB for hard-decision, 8-level soft-decision, or infinite-level quantization, respectively. This also predicts a coding gain of approximately 1.5 dB by switching from hard-decision to soft-decision decoding. Unlike uncoded BPSK, which requires 9.6 dB to achieve an error probability of 10^{-5}, near-error-free communication is possible with infinite-level quantization at an SNR of 0.19 dB using BPSK over an AWGN channel with a rate 1/2 error-control code. We have come very close to achieving this performance with rate 1/2 Polar, LDPC, and Turbo codes, as we shall explore in the chapters to follow.

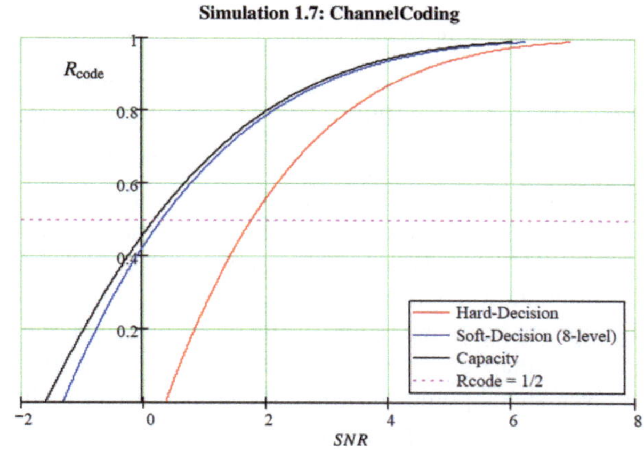

Fig. 1.16 The minimum SNR required for near-error-free communication using BPSK over an AWGN channel with a rate R_{code} error-control code

1.5 Problems

1. Consider a DMC that consists of the serial concatenation of two BSCs, each with crossover probability α. You may assume $P(x_1) = P(x_2) = 0.5$. Show that this concatenation is equivalent to a single BSC with a crossover probability given by $\beta = 2\alpha(1-\alpha)$.
2. Suppose the binary digits 0 and 1 are represented by $s_1(t) = A\cos(2\pi f_c t)$ and $s_2(t) = -A\cos(2\pi f_c t)$, respectively, where A is the amplitude, and f_c is the frequency of the sinusoidal carrier signal. (i) Show that the energy of a BPSK signal is $E_b = \frac{A^2 T_b}{2}$, where T_b is the time duration of the signal. (ii) Show that for a noiseless channel in which $r(t)$ is either $s_1(t)$ or $s_2(t)$, the signal point coordinate x_h is either $\sqrt{E_b}$ or $-\sqrt{E_b}$, respectively.
3. For uncoded BPSK with hard-decision decoding, show that $P(0|1) = Q\left(\sqrt{\frac{2E_b}{N_o}}\right)$.
4. Show that the QPSK signal will have twice the energy of a BPSK signal if the amplitude A and frequency f_c are the same as those of a BPSK signal.
5. In BPSK, if the basis function is given by $\phi_1(t) = \sqrt{\frac{2}{T_b}}\cos(2\pi f_c t + \theta)$, determine the BSC crossover probability α. Make use of $\int_0^{T_b} \cos(2\pi f_c t)\cos(2\pi f_c t + \theta)\,dt = \frac{T_b \cos\theta}{2}$. What value of θ will make $\alpha = 1/2$?
6. A QPSK signal is given by $s_i(t) = \sqrt{\frac{2E_s}{T_s}}\cos\left(2\pi f_c t + (2i-1)\frac{\pi}{4}\right)$ where $i = 1, 2, 3, 4$. The basis functions for the coherent QPSK demodulator are given by $\phi_1(t) = \sqrt{\frac{2}{T_s}}\cos(2\pi f_c t)$ and $\phi_2(t) = \sqrt{\frac{2}{T_s}}\sin(2\pi f_c t)$. (i) For $i = 1$, show that $s_1(t) = \sqrt{\frac{E_s}{2}}\,[\phi_1(t) - \phi_2(t)]$. Hence, show that the coordinates of

this signal point under a noiseless channel are $\left(\sqrt{\frac{E_s}{2}}, -\sqrt{\frac{E_s}{2}}\right)$. (ii) Plot this corresponding signal point and show that it lies on a radius of $\sqrt{E_s}$.

7. In M-ary Phase Shift Keying (PSK), $\log_2 M$ code digits are represented by a signal $s_i(t)$ of time duration T_s given by $s_i(t) = A\cos\left(2\pi f_c t + \frac{2\pi i}{M}\right)$, where f_c is the carrier frequency, and A is the amplitude of the signal. (i) Show that for no reduction in information throughput, the time duration of an MPSK signal $T_s = R_{\text{code}} T_b \log_2 M$ seconds and its energy $E_s = R_{\text{code}} E_b \log_2 M$. (ii) Sketch the signal constellation diagram for $M = 4$ for $R_{\text{code}} = 1$ and for $R_{\text{code}} = 1/2$, side by side. Indicate the radius of the circle on which the points lie and the coordinates of the signal points. (iii) By using the probability of correctly receiving a given signal point, show that the probability of a binary digit error $P_e = \frac{2\alpha - \alpha^2}{2}$, where $\alpha = Q\left(\sqrt{\frac{2E_b}{N_o}}\right)$ is the BSC crossover probability under uncoded BPSK.
8. Show that the probability of an error in 8-PSK for a demodulated code digit is worse than BPSK due to the reduced distance between signal constellation points.
9. The probability of an error in a symbol P_s for uncoded M-ary PSK is approximately given by $P_s = 2Q\left(\sqrt{\frac{2E_s}{N_o}}\sin\left(\frac{\pi}{M}\right)\right)$, where $Q(u)$ may be approximated by $Q(u) \approx \frac{1}{u\sqrt{2\pi}}\exp\left(-\frac{u^2}{2}\right)$. For $M = 16$ and $r_b = 1000$ binary digits/sec, the average carrier amplitude at the receiver is measured to be 1.5 mV. Assuming the single-sided noise power spectral density $N_o = 1 \times 10^{-11}$ W/Hz, determine the average number of binary errors made per second by the demodulator. Assume the use of gray coding.
10. Using a rate $1/N$ repetition code over a BSC, show that if the information throughput is allowed to be reduced, then the P_e at the sink is given by $P_{rep}(N, \alpha) = \sum_{i=\frac{(N-1)}{2}+1}^{N} \frac{N!}{(N-i)!i!}\alpha^i (1-\alpha)^{N-i}$.
11. The average mutual information for a DMC is given by $I(X, Y) = H(Y) - H(Y|X) = \sum_j P(y_j)\log_2\frac{1}{P(y_j)} - \sum_i\sum_j P(y_j|x_i)p(x_i)\log_2\frac{1}{P(y_j|x_i)}$ bits/symbol. Show that the channel capacity of a BSC is given by $C_s = [1 - \Omega(\alpha)]$ bits/symbol, where C_s is the maximum value of $I(X, Y)$. State any necessary assumptions.
12. Consider a DMC in which the input is binary and the output is either a binary digit 0, 1, or E. The forward transition probability $P(y_1|x_1) = P(y_3|x_2) = p$ and $P(y_2|x_1) = P(y_2|x_2) = 1-p$. (i) Show that $H(Y) = \sum_j P(y_j)\log_2\frac{1}{P(y_j)}$ is $p + \Omega(p)$, where $\Omega(\cdot)$ is the binary entropy function. (ii) Given that $H(Y|X) = \Omega(p)$, what is the average mutual information $I(X, Y) = H(Y) - H(Y|X)$ for this DMC and its channel capacity C_s?
13. The average mutual information for a DMC is given by $I(X, Y) = H(Y) - H(Y|X) = \sum_j P(y_j)\log_2\frac{1}{P(y_j)} - \sum_i\sum_j P(y_j|x_i)p(x_i)\log_2\frac{1}{P(y_j|x_i)}$

bits/symbol. For a DMC in which $P(x_1) = p$, $P(x_2) = P(x_3)$ with forward transition probabilities $P(y_1|x_1) = 1$, $P(y_2|x_3) = P(y_3|x_2) = \alpha$, (i) Show that $H(Y) = \Omega(p) + 1 - p$, where $\Omega(\cdot)$ is the binary entropy function. (ii) Show that $H(Y|X) = (1-p)\Omega(\alpha)$ and hence determine an expression for the average mutual information $I(X, Y)$. (iii) Under what conditions will the average mutual information be equal to the source entropy? Verify your answer.

14. Consider the use of 4-level soft-decision on the BPSK demodulator, which outputs the symbol 0, 1, 2, or 3. The integers [0,1] represent a binary digit 0, and [2, 3] represent a binary digit 1. The separation between each quantization level is a. The probability density function that governs the transmission of a binary digit 0 is $f(x_h|0) = \frac{1}{\sigma\sqrt{2\pi}}e^{-\frac{(x+\sqrt{E_b})^2}{2\sigma^2}}$. (i) Show that $P(2|0) = Q\left(\frac{\sqrt{E_b}}{\sigma}\right) - Q\left(\frac{a+\sqrt{E_b}}{\sigma}\right)$. (ii) Now determine $P(3|0)$ and add this to $P(2|0)$. Comment on this result. (iii) For a SNR of 0dB, what is the BSC crossover probability? You may use the approximation $Q(u) \approx \frac{1}{u\sqrt{2\pi}}e^{-\frac{u^2}{2}}$ and assume $\sigma = \sqrt{\frac{N_o}{2}}$.
15. Determine the forward transition probabilities in Fig. 1.15 under an AWGN channel. The separation between each quantization level is a.

Chapter 2
Polar Codes

In nature, a complex organism is made up of the simplest atoms. Similarly, the encoder of a *polar* code consists of numerous nodes, each performing a simple operation. Surprisingly, the decoding algorithm is also straightforward. This simplicity, combined with the powerful error-correction capability of polar codes, makes them an ideal starting point for understanding modern error-control coding techniques. Polar codes [5], invented by Prof. Arikan in 2009, have been adopted in various practical applications, including the 5G wireless communication standard [6], satellite communications, and data storage systems such as solid-state drives (SSDs). Their low-complexity decoding and capacity-achieving performance also make them ideal for emerging applications, for example, IoT networks, autonomous vehicles, and industrial automation.

We begin with the operation of a single node in Sect. 2.1 and then consider a detailed step-by-step explanation of a (4, 2) polar code in which two message bits are encoded into four code digits in Sect. 2.2. The corresponding decoding algorithm is presented in detail with a working example in Sect. 2.3. Thereafter, in Sect. 2.4, a more complex (8, 4) polar code is presented with its corresponding decoder in Sect. 2.5 and its BER performance in Sect. 2.6. Section 2.7 delves further into the concept of the reliability sequence, and in Sect. 2.8, we discover why a theoretically more complex optimum function for the operation of a single node is unnecessary. The chapter concludes with appendices that present some of the mathematical derivations that might distract from the main flow of thought and provide the 5G standard reliability sequence.

J. Sodha, *Fundamentals of Polar, LDPC and Turbo Codes*,
https://doi.org/10.1007/978-3-031-90924-5_2

2.1 Fundamentals

Polar codes are block codes in which K message binary digits are input to an encoder, which outputs N code digits. We shall use the notation (N, K) to denote such a polar code with code rate $R_{code} = K/N$. To understand the fundamental building block of a given polar code, consider the nodes shown in Fig. 2.1 in which the parent node is connected to lower left-child and right-child nodes. The left node is labeled u_1, and the right node is labeled u_2 to represent the two input binary digits. The output of this node are two code digits labeled c_1 and c_2.

If we express the input word as $\mathbf{u} = [u_1, u_2]$, the corresponding codeword output is $\mathbf{c} = [c_1, c_2]$, where $c_1 = (u_1 \oplus u_2)$ and $c_2 = u_2$, where the "$\oplus$" sign denotes the modulo-2 operation (standard XOR logic gate) presented in Table 2.1.

Equivalently, using the *polar transform* generator matrix,

$$\mathbf{G} = \begin{bmatrix} 1 & 0 \\ 1 & 1 \end{bmatrix} \tag{2.1}$$

the codeword $\mathbf{c} = \mathbf{uG}$ is given by

$$\mathbf{c} = [u_1, u_2] \cdot \begin{bmatrix} 1 & 0 \\ 1 & 1 \end{bmatrix} = [u_1 \cdot 1 \oplus u_2 \cdot 1, u_2 \cdot 1] = [u_1 \oplus u_2, u_2] \tag{2.2}$$

Notice that the transform is invertible because $u_1 = c_1 \oplus c_2$. Let us assume that these code digits c_1 and c_2 are transmitted over an AWGN channel using BPSK in which a binary digit 0 is mapped to the signal point $\sqrt{E_s}$ and a binary digit 1 is mapped to $-\sqrt{E_s}$. To ensure $E_s = 1$ joule without any loss of generality,

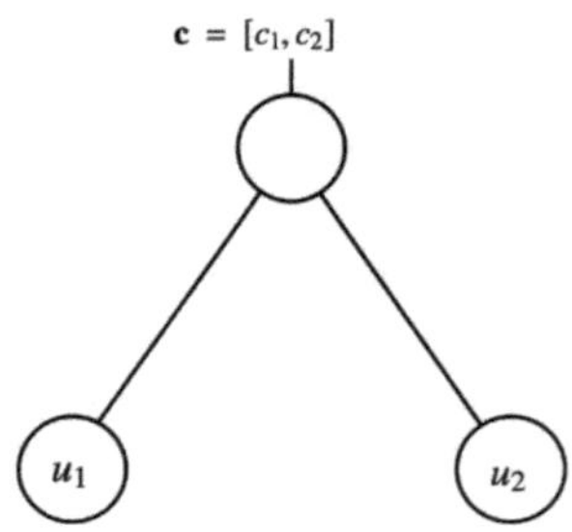

Fig. 2.1 Parent node connected to a left and a right node

Table 2.1 XOR logic

Modulo-2 addition		
x_1	x_2	$x_1 \oplus x_2$
0	0	0
0	1	1
1	0	1
1	1	0

Table 2.2 Belief analysis

u_1	u_2	c_1	c_2	r_1	r_2	sgn(r_1)	sgn(r_2)	Belief $g(r_1, r_2, u_1)$	sgn(r_1)sgn(r_2)	Belief $f(r_1, r_2)$
0	0	0	0	1	1	+	+	2	+	1
0	1	1	1	−1	−1	−	−	−2	+	1
1	0	1	0	−1	1	−	+	2	−	−1
1	1	0	1	1	−1	+	−	−2	−	−1

we shall set $E_b = 1/R_{code}$. Let $\mathbf{r} = [r_1, r_2]$ represent the corresponding received word in which r_1 and r_2 are the signal point coordinates along the horizontal axis of the BPSK signal constellation. From Eq. 2.2 if $u_1 = 0$, then $\mathbf{c} = [c_1, c_2] = [0 \oplus u_2, u_2] = [u_2, u_2]$. Namely, $c_1 = c_2 = u_2$, and accordingly, the received word is a repetition code. Specifically, since $[r_1, r_2]$ corresponds to $[c_1, c_2]$, the *belief* for u_2 can be taken to be $(r_1 + r_2)$. This belief is a measure of confidence. Conversely if $u_1 = 1$, then $c_1 = \overline{u_2}$ (i.e., its inverse) and $c_2 = u_2$. Once again this is a repetition code in which the belief for u_2 is $(-1)r_1 + r_2 = (r_2 - r_1)$. Therefore, we define the following function for the belief on u_2

$$g(a, b, i) = b + (1 - 2i)a \tag{2.3}$$

where a and b are the incoming beliefs to this node. In our example, these are r_1 and r_2, respectively. However, these received values should ideally be scaled as shown in problem 14. In problem 15, Eq. 2.3 is derived formally. Note that $g(a, b, 0) = b+a$ and $g(a, b, 1) = b-a$. Let us now consider the belief on u_1. From Eq. 2.2, given that $u_1 = c_1 \oplus c_2$, the multiplication of the signs $(sign(r_1)sign(r_2))$ is the hard-decisioned result of $r_1 \oplus r_2$. This is illustrated in Table 2.2 in which a noiseless channel is assumed and all possible input combinations to the node are explored. Bear in mind also that a binary digit 0 corresponds to a BPSK signal point with a positive coordinate. If the multiplication $(sign(r_1)sign(r_2))$ is positive, this corresponds to $u_1 = 0$. Conversely if the multiplication is negative, then $u_1 = 1$ as expected. Using the intuitive logic that a chain is only as strong as its weakest link, we may take the belief on u_1 to be given by the following function:

$$f(a, b) = sign(a)sign(b)min(|a|, |b|) \tag{2.4}$$

where $min(|a|, |b|)$ is minimum value of the two magnitudes. For this simple noiseless example, the $min(|r_1|, |r_2|) = 1$ and the corresponding beliefs on u_1 are listed in the last column of Table 2.2. Notice that $g(a, b, i)$ depends on the value of i. Therefore, the $f(.)$ function is used first to determine the left-child belief. This belief is then hard-decisioned to find the value of i, and then the $g(.)$ function is used to determine the right-child belief.

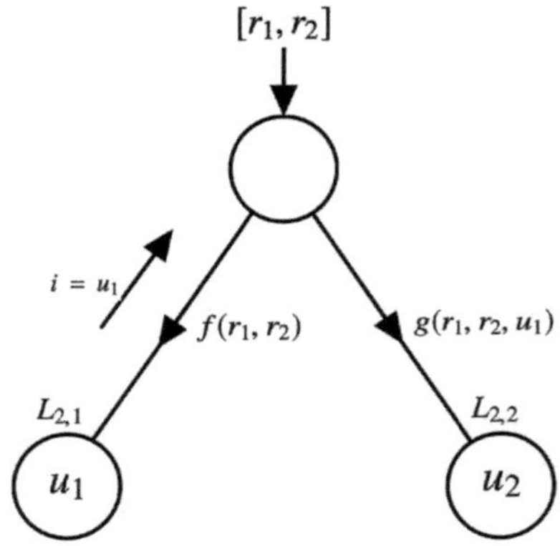

Fig. 2.2 Decoding left- and right-child nodes

To summarize, the operation of a single node within a polar decoder is presented in Fig. 2.2 in which the received word $[r_1, r_2]$ is assigned to the parent node. The output of the functions $f(.)$ and $g(.)$ are beliefs that are presented in a *belief matrix* $\mathbf{L}$ given by

$$\mathbf{L} = \begin{bmatrix} L_{1,1} & L_{1,2} \\ L_{2,1} & L_{2,2} \end{bmatrix} \tag{2.5}$$

For example if $\mathbf{c} = [1, 1]$ is transmitted using BPSK over an AWGN channel and received as $\mathbf{r} = [r_1, r_2] = [-1.23, -0.45]$, then $L_{1,1} = r_1 = -1.23$ and $L_{1,2} = r_2 = -0.45$. Using the $f(.)$ function, the belief for u_1 is $L_{2,1} = f(-1.23, -0.45) = (-1)(-1)0.45 = 0.45$. This is now hard-decisioned to find the value of i within the $g(.)$ function as follows. Given that a binary digit 0 corresponds to a positive belief in our BPSK modulation scheme, let us define the following function to hard-decision a given belief $L_{i,j}$

$$Decision(L_{i,j}) = \begin{cases} 0 & \text{if } L_{i,j} \geq 0, \\ 1 & \text{otherwise.} \end{cases} \tag{2.6}$$

Thus, we decode $u_1 = Decision(L_{2,1}) = Decision(0.45) = 0$. Now using the $g(.)$ function with $i = u_1 = 0$, we have $g(a, b, i) = g(-1.23, -0.45, 0) = r_2 + r_1 = -0.45 - 1.23 = -1.68 = L_{2,2}$. Therefore, $u_2 = Decision(-1.68) = 1$ to yield the output $\mathbf{u} = [u_1, u_2] = [0, 1]$.

The powerful error-control capability of polar codes stems from the inclusion of *frozen* binary digits, which are binary digit 0s, within the input word $\mathbf{u} = [u_1, u_2]$ to reduce error propagation. This concept is presented next.

2.2 (4, 2) Polar Code Encoder

Having introduced the underlying structure and notation, consider the (4, 2) polar code presented in Fig. 2.3. In this example, the nodes labeled u_1, u_2, u_3, u_4 at the bottom of the tree contain two message binary digits m_1, m_2 located at the nodes u_3, u_4, and frozen binary digits values of 0 are assigned to the nodes u_1, u_2. The selection of which nodes are assigned to be the frozen will be considered in Sect. 2.7. For now assume that the frozen and message binary digit positions are as presented in Fig. 2.3. Starting with the bottom left and right nodes u_1 and u_2, the output of the corresponding parent node is denoted by $[a, b]$, where $a = u_1 \oplus u_2$ and $b = u_2$. Listing the output of all the parent nodes using the letters of the alphabet, we find

$$\begin{aligned} [a, b] &= [u_1 \oplus u_2, u_2] \\ [c, d] &= [u_3 \oplus u_4, u_4] \end{aligned} \tag{2.7}$$

which combine to produce the output of the uppermost parent $[c_1, c_2, c_3, c_4]$ given by

$$[c_1, c_2, c_3, c_4] = [a \oplus c, b \oplus d, c, d] = [u_1 \oplus u_2 \oplus u_3 \oplus u_4, u_2 \oplus u_4, u_3 \oplus u_4, u_4] \tag{2.8}$$

Equivalently, the corresponding generator matrix is given by the Kronecker product

$$\mathbf{G}^{\otimes 2} = \mathbf{G} \otimes \mathbf{G} = \begin{bmatrix} 1 & 0 \\ 1 & 1 \end{bmatrix} \otimes \begin{bmatrix} 1 & 0 \\ 1 & 1 \end{bmatrix} = \begin{bmatrix} 1 \cdot \begin{bmatrix} 1 & 0 \\ 1 & 1 \end{bmatrix} & 0 \cdot \begin{bmatrix} 1 & 0 \\ 1 & 1 \end{bmatrix} \\ 1 \cdot \begin{bmatrix} 1 & 0 \\ 1 & 1 \end{bmatrix} & 1 \cdot \begin{bmatrix} 1 & 0 \\ 1 & 1 \end{bmatrix} \end{bmatrix} = \begin{bmatrix} 1 & 0 & 0 & 0 \\ 1 & 1 & 0 & 0 \\ 1 & 0 & 1 & 0 \\ 1 & 1 & 1 & 1 \end{bmatrix} \tag{2.9}$$

Fig. 2.3 (4, 2) polar code

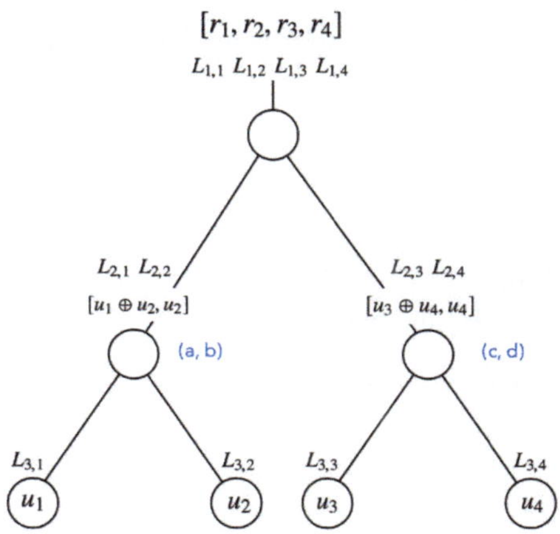

Table 2.3 (4,2) polar codewords.

m_1	m_2	c_1	c_2	c_3	c_4
0	0	0	0	0	0
0	1	1	1	1	1
1	0	1	0	1	0
1	1	0	1	0	1

and the codeword $\mathbf{c} = \mathbf{u}\mathbf{G}^{\otimes 2}$ is given by

$$[u_1, u_2, u_3, u_4] \cdot \begin{bmatrix} 1\,0\,0\,0 \\ 1\,1\,0\,0 \\ 1\,0\,1\,0 \\ 1\,1\,1\,1 \end{bmatrix} = [u_1 \oplus u_2 \oplus u_3 \oplus u_4,\, u_2 \oplus u_4,\, u_3 \oplus u_4,\, u_4] \tag{2.10}$$

If we assign the frozen bits to $u_1 = u_2 = 0$ and the message bits to $u_3 = m_1, u_4 = m_2$, the code digits are given by

$$c_1 = m_1 \oplus m_2$$
$$c_2 = m_2$$
$$c_3 = m_1 \oplus m_2$$
$$c_4 = m_2$$

Table 2.3 lists all the possible input combinations and corresponding codewords for this (4, 2) polar code. In the next section, we shall consider the decoder for this (4, 2) code.

2.3 (4, 2) Polar Code Decoder

Let $[r_1, r_2, r_3, r_4]$ represent the received BPSK signal points corresponding to the transmitted codeword $[c_1, c_2, c_3, c_4]$. We begin by defining the belief matrix $\mathbf{L}$ and follow steps 1–7 to decode the message bits m_1 and m_2.

$$\mathbf{L} = \begin{bmatrix} L_{1,1} & L_{1,2} & L_{1,3} & L_{1,4} \\ L_{2,1} & L_{2,2} & L_{2,3} & L_{2,4} \\ L_{3,1} & L_{3,2} & L_{3,3} & L_{3,4} \end{bmatrix} \tag{2.11}$$

Step 1
Initialize the beliefs .

$$L_{1,1} = r_1,\, L_{1,2} = r_2,\, L_{1,3} = r_3,\, L_{1,4} = r_4$$

Step 2
Determine the left nodes one level below

$$L_{2,1} = f(L_{1,1}, L_{1,3})$$
$$L_{2,2} = f(L_{1,2}, L_{1,4})$$

Step 3
Determine the left-child belief $L_{3,1}$. Even though the belief $L_{3,1}$ is found in this step, it is not used because it is a frozen node, and accordingly, $u_1 = 0$.

$$L_{3,1} = f(L_{2,1}, L_{2,2})$$
$$u_1 = 0$$

Step 4
Determine the right-child $L_{3,2}$ belief using the decision on $L_{3,1}$. The belief $L_{3,2}$ is not used because the second node u_2 is a frozen bit. Once again we simply assign the value of $u_2 = 0$.

$$L_{3,2} = g(L_{2,1}, L_{2,2}, u_1) = g(L_{2,1}, L_{2,2}, 0)$$
$$u_2 = 0$$

Step 5

$$L_{2,3} = g(L_{1,1}, L_{1,3}, u_1 \oplus u_2)$$
$$L_{2,4} = g(L_{1,2}, L_{1,4}, u_2)$$

Step 6
Decode the first message bit $m_1 = u_3$.

$$L_{3,3} = f(L_{2,3}, L_{2,4})$$
$$\hat{m}_1 = u_3 = Decision(L_{3,3})$$

Step 7
Decode the second message bit $m_2 = u_4$.

$$L_{3,4} = g(L_{2,3}, L_{2,4}, u_3)$$
$$\hat{m}_2 = u_4 = Decision(L_{3,4})$$

The decoding algorithm presented here is referred to as the *Successive Cancellation* (SC) decoding algorithm. Notice how the decoding algorithm systematically traverses through the polar tree from the top toward the bottom left most node, then to its corresponding right child before moving upward and downward as required through all the nodes. The bits are decoded one by one in a sequential manner, starting from the first bit and moving to the last bit. The beliefs (or likelihoods) are propagated through the decoding tree and used to make decisions.

2.3.1 Example: (4, 2) Polar Code over an AWGN via BPSK

In our communication system, we shall assume the code digits are transmitted over an AWGN channel using BPSK in which a binary digit 0 is mapped to the signal point $\sqrt{E_s}$ and a binary digit 1 is mapped to $-\sqrt{E_s}$ where $E_s = E_b R_{code} = 1$ joule by setting $E_b = 1/R_{code}$. For $m_1 = m_2 = 1$, the corresponding codeword is $[c_1, c_2, c_3, c_4] = [0101]$. Ideally, under a noiseless channel, the received word would be [1, -1, 1, -1]. Under noisy conditions, let the corresponding received signal points be $[r_1, r_2, r_3, r_4] = [1.226, -0.645, 0.854, -1.154]$. Following steps 1–7, the belief matrix would evolve as follows.

Step 1

$$\mathbf{L} = \begin{bmatrix} 1.226 & -0.645 & 0.854 & -1.154 \\ 0 & 0 & 0 & 0 \\ 0 & 0 & 0 & 0 \end{bmatrix}$$

Step 2

$$L_{2,1} = f(1.226, 0.854) = 0.854$$
$$L_{2,2} = f(-0.645, -1.154) = 0.645$$

$$\mathbf{L} = \begin{bmatrix} 1.226 & -0.645 & 0.854 & -1.154 \\ 0.854 & 0.645 & 0 & 0 \\ 0 & 0 & 0 & 0 \end{bmatrix}$$

Step 3

$$L_{3,1} = f(0.854, 0.645) = 0.645$$
$$u_1 = 0$$

$$\mathbf{L} = \begin{bmatrix} 1.226 & -0.645 & 0.854 & -1.154 \\ 0.854 & 0.645 & 0 & 0 \\ 0.645 & 0 & 0 & 0 \end{bmatrix}$$

Step 4

$$L_{3,2} = g(0.854, 0.645, 0) = (0.854 + 0.645) = 1.499$$
$$u_2 = 0$$

$$\mathbf{L} = \begin{bmatrix} 1.226 & -0.645 & 0.854 & -1.154 \\ 0.854 & 0.645 & 0 & 0 \\ 0.645 & 1.499 & 0 & 0 \end{bmatrix}$$

Step 5

$$L_{2,3} = g(1.226, 0.854, 0) = 1.226 + 0.854 = 2.08$$
$$L_{2,4} = g(-0.645, -1.154, 0) = -0.645 - 1.154 = -1.799$$

$$\mathbf{L} = \begin{bmatrix} 1.226 & -0.645 & 0.854 & -1.154 \\ 0.854 & 0.645 & 2.08 & -1.799 \\ 0.645 & 1.499 & 0 & 0 \end{bmatrix}$$

Step 6

$L_{3,3} = f(2.08, -1.799) = -1.799$
$u_3 = Decision(-1.799) = 1$

$$\mathbf{L} = \begin{bmatrix} 1.226 & -0.645 & 0.854 & -1.154 \\ 0.854 & 0.645 & 2.08 & -1.799 \\ 0.645 & 1.499 & -1.799 & 0 \end{bmatrix}$$

Step 7

$L_{3,4} = g(2.08, -1.799, 1) = -1.799 - 2.08 = -3.879$
$u_4 = Decision(-3.879) = 1$

$$\mathbf{L} = \begin{bmatrix} 1.226 & -0.645 & 0.854 & -1.154 \\ 0.854 & 0.645 & 2.08 & -1.799 \\ 0.645 & 1.499 & -1.799 & -3.879 \end{bmatrix}$$

The decoded message bits $[\hat{m}_1, \hat{m}_2] = [u_3, u_4] = [1, 1]$. Notice how the belief values $L_{3,j}$ increase in magnitude from left to right for $j = 1, 2, 3, 4$. This *polarization* effect causes an **unequal** level of error protection on the message bits. Specifically in this example, m_2 has a higher level of protection than m_1. This is demonstrated by simulation next.

2.3.2 *Simulation: (4, 2) Polar Code Performance*

In simulation-N4BER, the probability of an error in a binary digit P_e performance of the (4, 2) polar code under an AWGN channel is presented in Fig. 2.4 as the channel SNR is varied from 0 dB to 10dB. Notice how P_e on m_2 is lower than on m_1 for a given SNR. We shall consider this in more detail in Sect. 2.7. Since neither curve provides a coding gain over the BPSK curve, this polar code would not be used in practice. A more useful polar code is considered next.

2.4 (8, 4) Polar Code Encoder

Increasing the level of complexity a little further, consider the (8, 4) polar code presented in Fig. 2.5, in which the nodes labeled $u_1, u_2, \ldots, u_8$ at the bottom of the tree contain four message binary digits m_1, m_2, m_3, m_4 located at the nodes u_4, u_6, u_7, u_8. The remaining nodes are assigned the frozen binary digits values of 0. Once again consider the left and right nodes u_1 and u_2. The output of its parent node is labeled $[a, b]$, where $a = u_1 \oplus u_2$ and $b = u_2$. Listing all the parent nodes in which the letters of the alphabet are used in order, we find

$[a, b] = [u_1 \oplus u_2, u_2]$
$[c, d] = [u_3 \oplus u_4, u_4]$

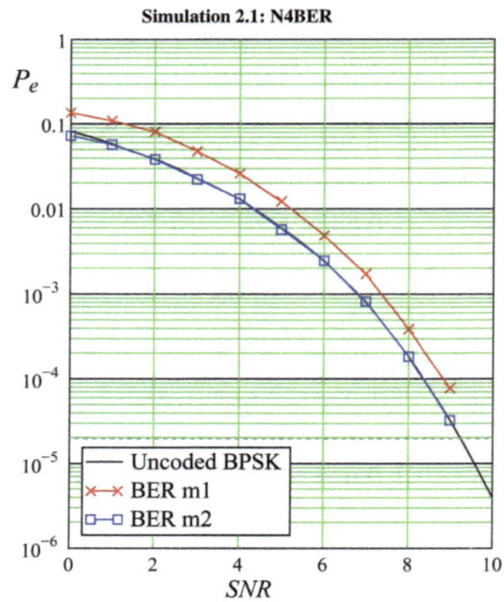

Fig. 2.4 BER Performance of the (4, 2) polar code

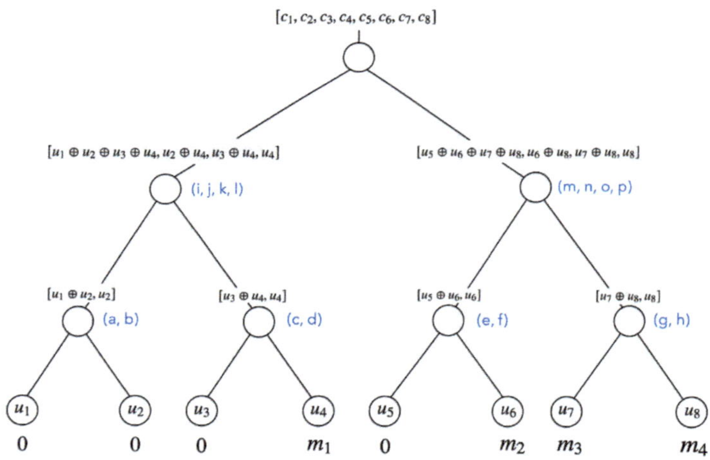

Fig. 2.5 (8, 4) polar code

$$[e, f] = [u_5 \oplus u_6, u_6]$$
$$[g, h] = [u_7 \oplus u_8, u_8]$$

Similarly, the parents of these nodes are given by

$$[i, j, k, l] = [a \oplus c, b \oplus d, c, d] = [u_1 \oplus u_2 \oplus u_3 \oplus u_4, u_2 \oplus u_4, u_3 \oplus u_4, u_4]$$
$$[m, n, o, p] = [e \oplus g, f \oplus h, g, h] = [u_5 \oplus u_6 \oplus u_7 \oplus u_8, u_6 \oplus u_8, u_7 \oplus u_8, u_8]$$

which combine to the uppermost parent

$$[c_1, c_2, c_3, c_4, c_5, c_6, c_7, c_8] = [i \oplus m, j \oplus n, k \oplus o, l \oplus p, m, n, o, p]$$

to give the following equations for the output code digits:

$$
\begin{aligned}
c_1 &= u_1 \oplus u_2 \oplus u_3 \oplus u_4 \oplus u_5 \oplus u_6 \oplus u_7 \oplus u_8\\
c_2 &= u_2 \oplus u_4 \oplus u_6 \oplus u_8\\
c_3 &= u_3 \oplus u_4 \oplus u_7 \oplus u_8\\
c_4 &= u_4 \oplus u_8\\
c_5 &= u_5 \oplus u_6 \oplus u_7 \oplus u_8\\
c_6 &= u_6 \oplus u_8\\
c_7 &= u_7 \oplus u_8\\
c_8 &= u_8
\end{aligned}
$$

Equivalently, the generator matrix is the Kronecker product $\mathbf{G}^{\otimes log_2 N} = \mathbf{G}^{\otimes 3}$ given by

$$
\mathbf{G}^{\otimes 3} = \mathbf{G}^{\otimes 2} \otimes \mathbf{G} = \begin{bmatrix} 1\,0\,0\,0 \\ 1\,1\,0\,0 \\ 1\,0\,1\,0 \\ 1\,1\,1\,1 \end{bmatrix} \otimes \begin{bmatrix} 1\,0 \\ 1\,1 \end{bmatrix} = \begin{bmatrix}
1\cdot\begin{bmatrix} 1\,0 \\ 1\,1 \end{bmatrix} & 0\cdot\begin{bmatrix} 1\,0 \\ 1\,1 \end{bmatrix} & 0\cdot\begin{bmatrix} 1\,0 \\ 1\,1 \end{bmatrix} & 0\cdot\begin{bmatrix} 1\,0 \\ 1\,1 \end{bmatrix} \\
1\cdot\begin{bmatrix} 1\,0 \\ 1\,1 \end{bmatrix} & 1\cdot\begin{bmatrix} 1\,0 \\ 1\,1 \end{bmatrix} & 0\cdot\begin{bmatrix} 1\,0 \\ 1\,1 \end{bmatrix} & 0\cdot\begin{bmatrix} 1\,0 \\ 1\,1 \end{bmatrix} \\
1\cdot\begin{bmatrix} 1\,0 \\ 1\,1 \end{bmatrix} & 0\cdot\begin{bmatrix} 1\,0 \\ 1\,1 \end{bmatrix} & 1\cdot\begin{bmatrix} 1\,0 \\ 1\,1 \end{bmatrix} & 0\cdot\begin{bmatrix} 1\,0 \\ 1\,1 \end{bmatrix} \\
1\cdot\begin{bmatrix} 1\,0 \\ 1\,1 \end{bmatrix} & 1\cdot\begin{bmatrix} 1\,0 \\ 1\,1 \end{bmatrix} & 1\cdot\begin{bmatrix} 1\,0 \\ 1\,1 \end{bmatrix} & 1\cdot\begin{bmatrix} 1\,0 \\ 1\,1 \end{bmatrix}
\end{bmatrix}
$$

$$
= \begin{bmatrix}
1\,0\,0\,0\,0\,0\,0\,0 \\
1\,1\,0\,0\,0\,0\,0\,0 \\
1\,0\,1\,0\,0\,0\,0\,0 \\
1\,1\,1\,1\,0\,0\,0\,0 \\
1\,0\,0\,0\,1\,0\,0\,0 \\
1\,1\,0\,0\,1\,1\,0\,0 \\
1\,0\,1\,0\,1\,0\,1\,0 \\
1\,1\,1\,1\,1\,1\,1\,1
\end{bmatrix} \tag{2.12}
$$

and the codeword $\mathbf{c} = \mathbf{u}\mathbf{G}^{\otimes 3}$ is given by

$$
\mathbf{c} = [u_1, u_2, u_3, u_4, u_5, u_6, u_7, u_8] \cdot \begin{bmatrix}
1\,0\,0\,0\,0\,0\,0\,0 \\
1\,1\,0\,0\,0\,0\,0\,0 \\
1\,0\,1\,0\,0\,0\,0\,0 \\
1\,1\,1\,1\,0\,0\,0\,0 \\
1\,0\,0\,0\,1\,0\,0\,0 \\
1\,1\,0\,0\,1\,1\,0\,0 \\
1\,0\,1\,0\,1\,0\,1\,0 \\
1\,1\,1\,1\,1\,1\,1\,1
\end{bmatrix} \tag{2.13}
$$

$$\mathbf{c}^T = \begin{bmatrix} c_1 \\ c_2 \\ c_3 \\ c_4 \\ c_5 \\ c_6 \\ c_7 \\ c_8 \end{bmatrix} = \begin{bmatrix} u_1 \oplus u_2 \oplus u_3 \oplus u_4 \oplus u_5 \oplus u_6 \oplus u_7 \oplus u_8 \\ u_2 \oplus u_4 \oplus u_6 \oplus u_8 \\ u_3 \oplus u_4 \oplus u_7 \oplus u_8 \\ u_4 \oplus u_8 \\ u_5 \oplus u_6 \oplus u_7 \oplus u_8 \\ u_6 \oplus u_8 \\ u_7 \oplus u_8 \\ u_8 \end{bmatrix} \tag{2.14}$$

Having assigned the frozen bits $u_1 = u_2 = u_3 = u_5 = 0$ and the message bits $u_4 = m_1, u_6 = m_2, u_7 = m_3, u_8 = m_4$, the code digits in terms of the message bits are given by

$$\begin{aligned}
c_1 &= 0 \oplus 0 \oplus 0 \oplus m_1 \oplus 0 \oplus m_2 \oplus m_3 \oplus m_4 = m_1 \oplus m_2 \oplus m_3 \oplus m_4 \\
c_2 &= 0 \oplus m_1 \oplus m_2 \oplus m_4 = m_1 \oplus m_2 \oplus m_4 \\
c_3 &= 0 \oplus m_1 \oplus m_3 \oplus m_4 = m_1 \oplus m_3 \oplus m_4 \\
c_4 &= m_1 \oplus m_4 \\
c_5 &= 0 \oplus m_2 \oplus m_3 \oplus m_4 = m_2 \oplus m_3 \oplus m_4 \\
c_6 &= m_2 \oplus m_4 \\
c_7 &= m_3 \oplus m_4 \\
c_8 &= m_4
\end{aligned}$$

2.5 (8, 4) Polar Code Decoder

To decode the received word $[r_1, r_2, r_3, r_4, r_5, r_6, r_7, r_8]$, we need to determine the following belief matrix:

$$\mathbf{L} = \begin{bmatrix} L_{1,1} & L_{1,2} & L_{1,3} & L_{1,4} & L_{1,5} & L_{1,6} & L_{1,7} & L_{1,8} \\ L_{2,1} & L_{2,2} & L_{2,3} & L_{2,4} & L_{2,5} & L_{2,6} & L_{2,7} & L_{2,8} \\ L_{3,1} & L_{3,2} & L_{3,3} & L_{3,4} & L_{3,5} & L_{3,6} & L_{3,7} & L_{3,8} \\ L_{4,1} & L_{4,2} & L_{4,3} & L_{4,4} & L_{4,5} & L_{4,6} & L_{4,7} & L_{4,8} \end{bmatrix} \tag{2.15}$$

where $L_{4,1}$ is the belief on node u_1, $L_{4,2}$ is the belief on node u_2, ..., $L_{4,8}$ is the belief on node u_8. Let $i = 1, 2, ..., k$ and $j = 1, 2, ..., n$. We begin by first assigning the beliefs $L_{1,j} = r_j$ for $j = 1$ to 8. To clarify this further, the polar code tree is shown again in Fig. 2.6, but now with the beliefs and message bits labeled.

The beliefs $L_{2,1}, L_{2,2}, L_{2,3}, L_{2,4}$ are calculated first and used to determine $L_{3,1}, L_{3,2}$, which in turn are used to determine the belief $L_{4,1}$ as follows.

Step 0

$L_{1,j} = r_j$ for $j = 1$ to 8

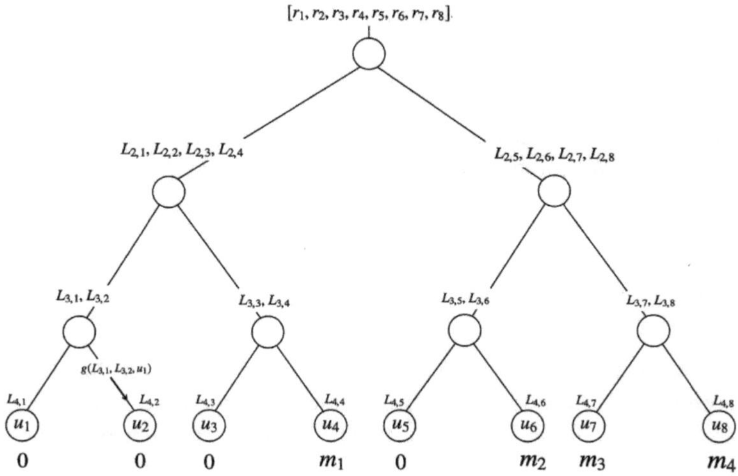

Fig. 2.6 (8, 4) polar code beliefs

Step 1

$$L_{2,1} = f(L_{1,1}, L_{1,5})$$
$$L_{2,2} = f(L_{1,2}, L_{1,6})$$
$$L_{2,3} = f(L_{1,3}, L_{1,7})$$
$$L_{2,4} = f(L_{1,4}, L_{1,8})$$

Step 2

$$L_{3,1} = f(L_{2,1}, L_{2,3})$$
$$L_{3,2} = f(L_{2,2}, L_{2,4})$$

Step 3

$$L_{4,1} = f(L_{3,1}, L_{3,2})$$
$$u_1 = 0$$

The belief $L_{4,1}$ is not hard-decisioned because it corresponds to a frozen binary digit. Thus, $u_1 = 0$ is assigned. The next step is to determine the corresponding right-child belief $L_{4,2}$ as follows.

Step 4

$$L_{4,2} = g(L_{3,1}, L_{3,2}, u_1) = g(L_{3,1}, L_{3,2}, 0)$$
$$u_2 = 0$$

Once again this node corresponds to a frozen binary digit, thus $u_2 = 0$.

Step 5

$$L_{3,3} = g(L_{2,1}, L_{2,3}, u_1 \oplus u_2) = g(L_{2,1}, L_{2,3}, 0)$$
$$L_{3,4} = g(L_{2,2}, L_{2,4}, u_2) = g(L_{2,2}, L_{2,4}, 0)$$

Step 6

$L_{4,3} = f(L_{3,3}, L_{3,4})$
$u_3 = 0$

The belief $L_{4,3}$ is not hard-decisioned because it corresponds to a frozen binary digit, thus $u_3 = 0$. We can now determine the corresponding right-child belief $L_{4,4}$ as follows.

Step 7

$L_{4,4} = g(L_{3,3}, L_{3,4}, u_3) = g(L_{3,3}, L_{3,4}, 0)$
$u_4 = Decision(L_{4,4})$

This is the first belief $L_{4,4}$ to be hard-decisioned (based on the BPSK signal points assignment) to decode $u_4 = \hat{m}_1$. Namely, the message symbol $\hat{m}_1 = 0$ if $L_{4,4} \geq 0$; otherwise, $\hat{m}_1 = 1$. Next, the belief $L_{4,5}$ is determined as follows.

Step 8

$L_{2,5} = g(L_{1,1}, L_{1,5}, u_1 \oplus u_2 \oplus u_3 \oplus u_4)$
$L_{2,6} = g(L_{1,2}, L_{1,6}, u_2 \oplus u_4)$
$L_{2,7} = g(L_{1,3}, L_{1,7}, u_3 \oplus u_4)$
$L_{2,8} = g(L_{1,4}, L_{1,8}, u_4)$
$L_{3,5} = f(L_{2,5}, L_{2,7})$
$L_{3,6} = f(L_{2,6}, L_{2,8})$
$L_{4,5} = f(L_{3,5}, L_{3,6})$
$u_5 = 0$

Given the belief $L_{4,5}$ corresponds to a frozen binary digit, we simply assign $u_5 = 0$. To determine the corresponding right-child belief $L_{4,6}$, the decision on $L_{4,5}$ is used as follows.

Step 9

$L_{4,6} = g(L_{3,5}, L_{3,6}, u_5) = g(L_{3,5}, L_{3,6}, 0)$
$u_6 = Decision(L_{4,6})$

This belief $L_{4,6}$ is hard-decisioned to decode u_6, which is the second message symbol $\hat{m}_2 = 0$ if $L_{4,6} \geq 0$; otherwise, $\hat{m}_2 = 1$.

Step 10

$L_{3,7} = g(L_{2,5}, L_{2,7}, u_5 \oplus u_6)$
$L_{3,8} = g(L_{2,6}, L_{2,8}, u_6)$
$L_{4,7} = f(L_{3,7}, L_{3,8})$
$u_7 = Decision(L_{4,7})$

This belief $L_{4,7}$ is hard-decisioned to decode u_7, which is the third message symbol $\hat{m}_3 = 0$ if $L_{4,7} \geq 0$; otherwise, $\hat{m}_3 = 1$.

Step 11

$L_{4,8} = g(L_{3,7}, L_{3,8}, u_7)$
$u_8 = Decision(L_{4,8})$

Finally, this belief $L_{4,8}$ is hard-decisioned to decode u_8, which is the fourth message symbol $\hat{m}_4 = 0$ if $L_{4,8} \geq 0$; otherwise, $\hat{m}_4 = 1$. Notice that for this (8, 4) polar code, the left-hand side of the tree below the upper most parent is used to decode only the message bit m_1. To appreciate this, consider the polar decoder steps for a noiseless channel. Given that $u_1 = u_2 = u_3 = 0$, we find

Step 0

$Decision(L_{1,j}) = c_j$ for $j = 1$ to 8

Step 1

$Decision(L_{2,1}) = c_1 \oplus c_5 = (u_1 \oplus u_2 \oplus u_3 \oplus u_4 \oplus u_5 \oplus u_6 \oplus u_7 \oplus u_8) \oplus (u_5 \oplus u_6 \oplus u_7 \oplus u_8) = u_4$
$Decision(L_{2,2}) = c_2 \oplus c_6 = (u_2 \oplus u_4 \oplus u_6 \oplus u_8) \oplus (u_6 \oplus u_8) = u_4$
$Decision(L_{2,3}) = c_3 \oplus c_7 = (u_3 \oplus u_4 \oplus u_7 \oplus u_8) \oplus (u_7 \oplus u_8) = u_4$
$Decision(L_{2,4}) = c_4 \oplus c_8 = (u_4 \oplus u_8) \oplus u_8 = u_4$

Steps 2, 3, and 4 are not required because u_1 and u_2 are frozen to a binary digit 0.

Step 5

$L_{3,3} = g(L_{2,1}, L_{2,3}, 0) = L_{2,1} + L_{2,3}$
$L_{3,4} = g(L_{2,2}, L_{2,4}, 0) = L_{2,2} + L_{2,4}$

Step 6 is not required because $u_3 = 0$.

Step 7

$L_{4,4} = g(L_{3,3}, L_{3,4}, 0) = L_{3,3} + L_{3,4} = (L_{2,1} + L_{2,3}) + (L_{2,2} + L_{2,4})$

In this noiseless case, the belief $L_{4,4} = \sum_{i=1}^{4} L_{2,i}$ before taking the decision $u_4 = Decision(L_{4,4})$. It may seem counterintuitive to use half a tree to make a decision on u_4 when the most reliable decoded bit is u_8 (refer to Sect. 2.7). But a correct decision on u_4 is critical to the overall BER performance of the polar code, as we shall discover in the next section. A more advanced list decoding algorithm for polar codes is presented in Appendix 2.10.

2.6 BER Performance of the (8, 4) Polar Code

The probability of an error in a binary digit P_e performance of the (8, 4) polar code under an AWGN channel is presented in Fig. 2.7a as the channel SNR is varied from 0 dB to 10 dB. The coding gain is approximately 1.5 dB for a BER of 10^{-5}.

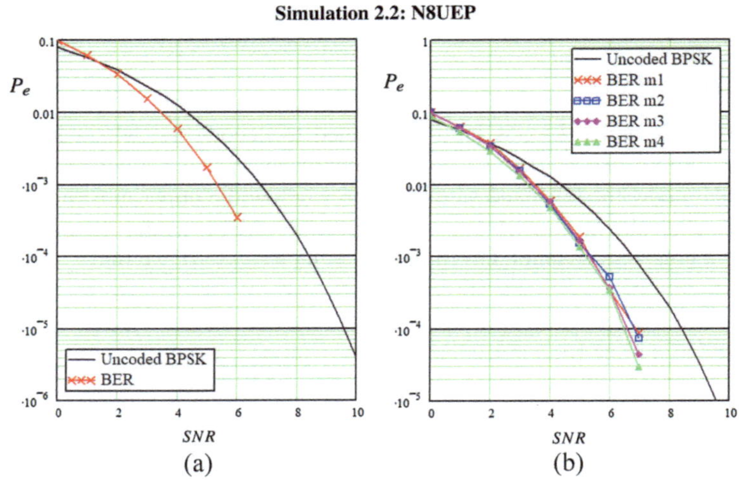

Fig. 2.7 (**a**) BER Performance of the (8, 4) polar code (**b**) Unequal error protection on individual bits

Fig. 2.8 BER Performance of the (8, 4) polar code assuming m_1 is known by the decoder

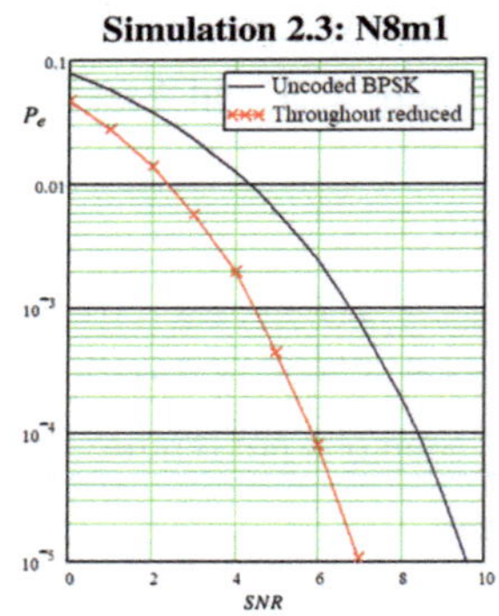

If we determine the probability of an error on each individual message bit, we find that the least reliable message bit is m_1, and the most reliable message bit is m_4 as shown in Fig. 2.7(b). This effect is more visible at higher SNRs. The average of these four BER curves is of course the BER curve presented in Fig. 2.7a. The unequal protection across the message bits stems from the inherent polarization effect of polar codes to be presented in the next section.

To demonstrate the importance of a correct decision on u_4, Fig. 2.8 shows the BER performance if the decoder correctly decodes u_4 always. The penalty incurred for this significant improvement in BER is a reduction in information throughput by 25%.

2.7 Reliability Sequence

In the previous section, it was shown that the level of error protection on a message bit depends on its location. The most reliable message bit positions can be identified using the *Bhattacharyya parameter* $Z(W)$, which measures the reliability of a communication channel W. Smaller values of $Z(W)$ indicate more reliable channels. In general, for a channel W with input alphabet $\mathcal{X}$ and output alphabet $\mathcal{Y}$, the Bhattacharyya parameter $Z(W)$ is defined as

$$Z(W) = \sum_{y \in \mathcal{Y}} \sqrt{P(y|0)P(y|1)} \tag{2.16}$$

where $P(y|x)$ is the probability of receiving y given the input x. For example, for a *Binary Erasure Channel* (BEC), each transmitted bit is either received correctly or erased with probability p. The output alphabet $\mathcal{Y}$ consists of three possible outcomes, $\{0, 1, \mathrm{e}\}$, where "e" denotes an erasure. The probabilities are $P(e|0) = p$, $P(e|1) = p$, $P(0|0) = 1 - p$, $P(1|1) = 1 - p$. Therefore, the Bhattacharyya parameter $Z(W)$ for the BEC is

$$Z(W) = \sqrt{P(e|0)P(e|1)} + \sqrt{P(0|0)P(0|1)} + \sqrt{P(1|0)P(1|1)} \tag{2.17}$$

Since $P(e|0) = P(e|1) = p$, $\sqrt{P(e|0)P(e|1)} = \sqrt{p \cdot p} = p$, $P(0|0) = 1 - p$, $P(1|1) = 1 - p$, $P(0|1) = P(1|0) = 0$, the Bhattacharyya parameter for a BEC simplifies to

$$Z(W) = p + 0 + 0 = p \tag{2.18}$$

Referring to the (8, 4) polar code tree in Fig. 2.9, the decoding process starts from the top parent node, which splits into left- and right-child nodes. Each split on the tree represents the creation of two *synthetic* channels. To determine the reliability sequence, the split is modeled as two synthetic BECs with different erasure probabilities. It is easily shown (problem 5) that the erasure probabilities of the left and right synthetic channels are $(2p - p^2)$ and (p^2), respectively.

Now for a perfectly reliable channel W, $Z(W) = 0$ and for a completely unreliable channel, $Z(W) = 1$. Thus, without any bias, we begin at the top of the tree with $Z(W) = 0.5$. One level down, the reliabilities of the left and right channels are given by $\left(2(0.5) - 0.5^2\right)$ and $\left(0.5^2\right)$, respectively, as shown in Fig. 2.9. At the base of the tree, the positions u_4, u_6, u_7, u_8 end up with the smallest values with the most reliable position being u_8 as confirmed by simulation in Fig. 2.7b. At the bottom of the tree, Table 2.4 lists the reliability values in descending order in which the bottom $K = 4$ positions corresponding to the message bits because these are the smallest values. The order $1, 2, 3, 5, 4, 6, 7, 8$ is referred to as the *reliability sequence*.

Simulation 2.4: Polarization

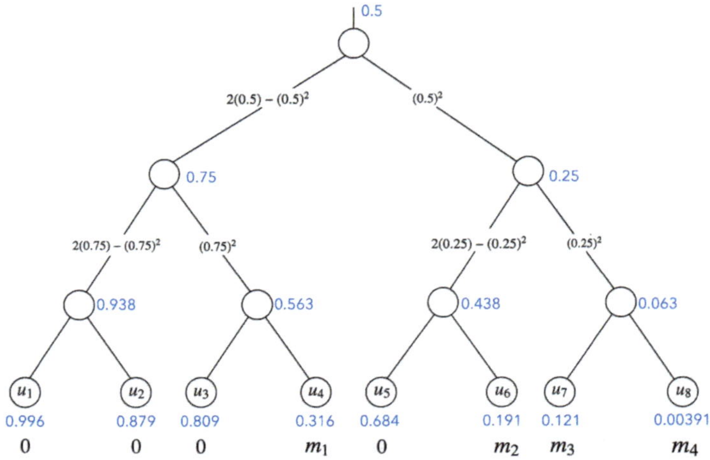

Fig. 2.9 (8, 4) polar code polarization

Table 2.4 (8, 4) polar code reliability sequence using $Z(W)$.

$Z(W)$	u_i
0.996	1
0.879	2
0.809	3
0.684	5
0.316	4
0.191	6
0.121	7
0.00391	8

In summary, even though there is physically only one communication channel, the original channel is effectively transformed into a set of synthetic channels, each with different reliability characteristics through the process of polarization. For example, for a (N = 16, K = 10) polar code, the reliability sequence is {1, 2, 3, 5, 9, 4, 6, 10, 7, 11, 13, 8, 12, 14, 15, 16} within which the frozen positions are {1, 2, 3, 5, 9, 4}. Note, however, that the method presented using $Z(W)$ calculations to determine the reliability sequence is only a simple first approximation for small values of N.

A more accurate method is to track the probability density functions of the log-likelihood ratios (LLRs) through the polar transform [5]. By undertaking this process, for $N = 1024$, the reliability sequence for the 5G wireless standard [6] is presented in Appendix 2.11. Using this, the reliability sequence for any value of $N \leq 1024$ can be determined by simply extracting the values less than or equal to N. For example, for $N = 16$, the numbers from this reliability sequence in order, including all the numbers from 1 to 16, are as follows:

1, 2, 3, 5, 9, 17, 33, 4, 6, 65, 10, 7, 18, 11, 19, 129, 13, 34, 66, 21, 257, 35, 25, 37, 8, 130, 67, 513, 12, 41, 69, 131, 20, 14, 49, 15, 73, 258, 22, 133, 36, 259, 27, 514, 81, 38, 26, 23, 137, 261, 265, 39, 515, 97, 68, 42, 145, 29, 70, 43, 517, 50, 75, 273, 161, 521, 289, 529, 193, 545, 71, 45, 132, 82, 51, 74, 16, 321, ...

Eliminating any number greater than 16 yields the reliability sequence 1, 2, 3, 5, 9, 4, 6, 10, 7, 11, 13, 8, 12, 14, 15, 16. The encoding and decoding of polar codes are now complete. The sections to follow consider the finer details of the decoding algorithm starting with the optimum $f(.)$ function.

2.8 Optimum f(.) Belief Function

In Sect. 2.1, the belief on the left child was determined via the low complexity function $f(a, b) = \text{sign}(a)\text{sign}(b)\min(|a|, |b|)$. In this section, we take a closer look at its origin from first principles to determine the optimum $f(.)$ function and present its influence on the BER performance of the (8, 4) polar code. Let us begin by considering the log-likelihood ratio (LLR) $\lambda(x)$, which provides a measure of confidence in the likelihood of x being 0 or 1.

$$\lambda(x) = \ln\left(\frac{P(x=0)}{P(x=1)}\right) \tag{2.19}$$

- If $\lambda(x)$ is large and positive, it indicates a high confidence (belief) that $x = 0$.
- If $\lambda(x)$ is large and negative, it indicates a high confidence (belief) that $x = 1$.
- If $\lambda(x)$ is close to 0, it indicates low confidence, meaning that x is equally likely to be 0 or 1.

The LLR function has all the properties required to develop an expression for the belief function $f(.)$ as follows.

Belief in $\mathbf{u_1}$

From Sect. 2.1, Eq. 2.2, given that $\mathbf{c} = [c_1, c_2] = [u_1 \oplus u_2, u_2]$, the belief in u_1 can be defined as

$$\lambda(c_1 \oplus c_2) = \ln\left(\frac{P(c_1 \oplus c_2 = 0)}{P(c_1 \oplus c_2 = 1)}\right) \tag{2.20}$$

where

$$P(c_1 \oplus c_2 = 0) = P(c_1 = 0)P(c_2 = 0) + P(c_1 = 1)P(c_2 = 1)$$

$$P(c_1 \oplus c_2 = 1) = P(c_1 = 0)P(c_2 = 1) + P(c_1 = 1)P(c_2 = 0)$$

For ease of notation, let $\lambda_1 = \lambda(c_1)$, so that

$$\lambda_1 = \ln\left(\frac{P(c_1 = 0)}{P(c_1 = 1)}\right) \tag{2.21}$$

Removing the ln(.) function

$$e^{\lambda_1} = \frac{P(c_1 = 0)}{P(c_1 = 1)} \tag{2.22}$$

Since c_1 must be either 0 or 1, $P(c_1 = 0) + P(c_1 = 1) = 1$

$$e^{\lambda_1} = \frac{1 - P(c_1 = 1)}{P(c_1 = 1)} \tag{2.23}$$

from which

$$P(c_1 = 1) = \frac{1}{1 + e^{\lambda_1}} \tag{2.24}$$

$$P(c_1 = 0) = \frac{e^{\lambda_1}}{1 + e^{\lambda_1}} \tag{2.25}$$

Similarly, let λ_2 represent $\lambda(c_2)$, we find

$$P(c_2 = 1) = \frac{1}{1 + e^{\lambda_2}} \tag{2.26}$$

$$P(c_2 = 0) = \frac{e^{\lambda_2}}{1 + e^{\lambda_2}} \tag{2.27}$$

Now, substituting these into the joint probabilities

$$P(c_1 \oplus c_2 = 0) = \frac{e^{\lambda_1}}{1 + e^{\lambda_1}} \cdot \frac{e^{\lambda_2}}{1 + e^{\lambda_2}} + \frac{1}{1 + e^{\lambda_1}} \cdot \frac{1}{1 + e^{\lambda_2}} \tag{2.28}$$

$$P(c_1 \oplus c_2 = 1) = \frac{e^{\lambda_1}}{1 + e^{\lambda_1}} \cdot \frac{1}{1 + e^{\lambda_2}} + \frac{1}{1 + e^{\lambda_1}} \cdot \frac{e^{\lambda_2}}{1 + e^{\lambda_2}} \tag{2.29}$$

Simplifying these, we get

$$P(c_1 \oplus c_2 = 0) = \frac{e^{\lambda_1 + \lambda_2} + 1}{(1 + e^{\lambda_1})(1 + e^{\lambda_2})} \tag{2.30}$$

$$P(c_1 \oplus c_2 = 1) = \frac{e^{\lambda_1} + e^{\lambda_2}}{(1 + e^{\lambda_1})(1 + e^{\lambda_2})} \tag{2.31}$$

Simulation 2.5: OptBelief

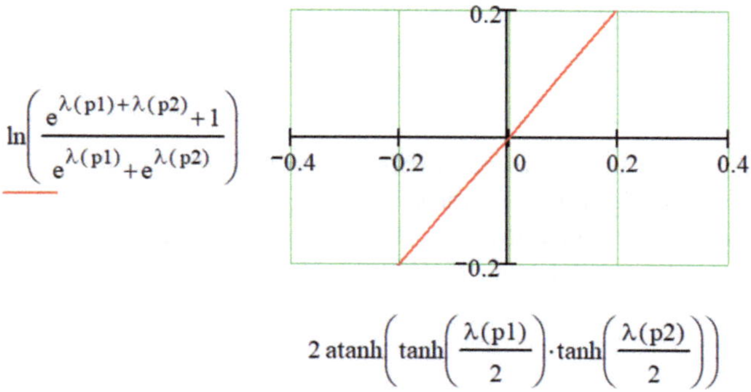

Fig. 2.10 Equivalent belief functions

Substituting these into Eq. 2.20, the log-likelihood of $(c_1 \oplus c_2)$ is given by

$$\lambda(c_1 \oplus c_2) = \ln\left(\frac{e^{\lambda_1+\lambda_2}+1}{e^{\lambda_1}+e^{\lambda_2}}\right) \tag{2.32}$$

Equation 2.32 is the optimum expression for the belief function $f(.)$. It is shown mathematically in Appendix 2.9 of this chapter that its equivalent to the following hyperbolic tangent expression, typically referred to as the *tanh-rule*:

$$f(\lambda_1, \lambda_2) = \lambda(c_1 \oplus c_2) = 2 \cdot \tanh^{-1}\left(\tanh\left(\frac{\lambda_1}{2}\right) \cdot \tanh\left(\frac{\lambda_2}{2}\right)\right) \tag{2.33}$$

Alternatively, Fig. 2.10 illustrates the equivalence of Eqs. 2.32 and 2.33 by defining $\lambda(p) = \ln\left(\frac{p}{1-p}\right)$ and setting for example, $p_1 = 0.4$ as p_2 ranges from 0 to 1. As expected, a plot of Eq. 2.32 versus Eq. 2.33 yields a line at 45 degrees. You may repeat this plot for other values of p_1.

2.8.1 AWGN Channel Posterior Probability

In Eq. 2.20, $\lambda(c_1 \oplus c_2)$ is the belief $f(.)$ on the modulo-2 addition of two code digits assuming the probabilities $P(c_1 = 0)$ and $P(c_2 = 0)$ are known. However, in a digital communication system, the code digits are transmitted over a channel. Let $\mathbf{c} = \{c_1, c_2, \cdots, c_n, \cdots, c_N\}$ represent the transmitted codeword and let $\mathbf{r} = \{r_1, r_2, \cdots, r_n, \cdots, r_N\}$ represent the corresponding received noisy word, where N

is the length of the codeword. The corresponding belief on the nth code digit is the log-likelihood ratio (LLR)

$$\lambda_n = \ln\left(\frac{P\left(c_n = 0|r_n\right)}{P\left(c_n = 1|r_n\right)}\right) \tag{2.34}$$

where $P\left(c_n = 0|r_n\right)$ is the *channel posterior* probability that the transmitted code digit c_n was a binary digit 0, given the received signal point coordinate r_n. Assuming that the code digits are transmitted via BPSK over an AWGN channel with a single-sided noise power spectral density N_o W/Hz with coherent demodulation, r_n is given by

$$r_n = \pm\sqrt{E_s} + n_n \tag{2.35}$$

where n_n is a *zero-mean Gaussian random variable* with standard deviation

$$\sigma = \sqrt{\frac{N_o}{2}} \tag{2.36}$$

To eliminate N_o from Eq. 2.36, we note that the channel signal-to-noise ratio SNR in decibels (dB) is given by

$$SNR = 10\log_{10}\left(\frac{E_b}{N_o}\right) \tag{2.37}$$

which substituted into Eq. 2.36 for N_o gives

$$\sigma = \sqrt{\frac{1}{2R_{code}\left(10^{\frac{SNR}{10}}\right)}} \tag{2.38}$$

where $E_b = \frac{1}{R_{code}}$ to ensure that the BPSK signal points are located at $\pm\sqrt{E_s} = \pm 1$. Using Bayes theorem

$$P(A|B) = \frac{P(B|A)P(A)}{P(B)} \tag{2.39}$$

the channel posterior probability $P\left(c_n = 0|r_n\right)$ in Eq. 2.34 is given by

$$P\left(c_n = 0|r_n\right) = \frac{P\left(r_n|c_n = 0\right)P\left(c_n = 0\right)}{P\left(r_n\right)} \tag{2.40}$$

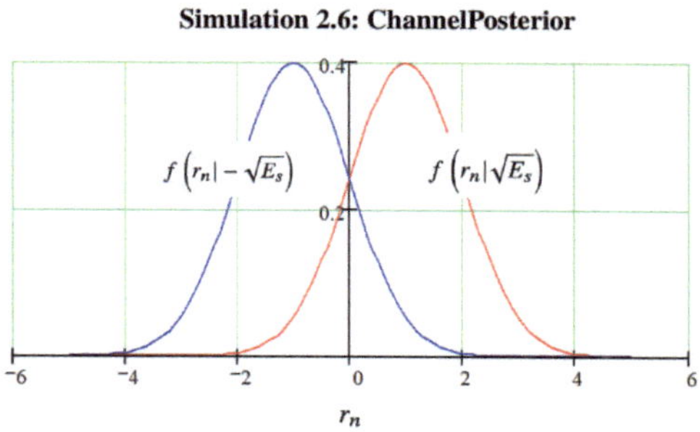

Fig. 2.11 Channel probability density functions for $E_s = 1$ joule

Equivalently (problem 6),

$$P\left(c_n = 0|r_n\right) = \frac{f\left(r_n|\sqrt{E_s}\right)P\left(c_n = 0\right)}{f\left(r_n\right)} \tag{2.41}$$

where the probability density functions (PDFs) are given by

$$f\left(r_n|\sqrt{E_s}\right) = \frac{1}{\sigma\sqrt{2\pi}}\exp\left[\frac{-\left(r_n - \sqrt{E_s}\right)^2}{2\sigma^2}\right] \tag{2.42}$$

$$f\left(r_n|-\sqrt{E_s}\right) = \frac{1}{\sigma\sqrt{2\pi}}\exp\left[\frac{-\left(r_n + \sqrt{E_s}\right)^2}{2\sigma^2}\right] \tag{2.43}$$

These PDFs are illustrated in Fig. 2.11 for $R_{code} = 1/2$, $E_b = 1/R_{code}$ at a channel $SNR = 0$ dB. The transition from Eq. 2.40 to 2.41 is verified via simulation in Sect. 2.8.2. Now combining the PDFs, we note that

$$f\left(r_n\right) = f\left(r_n|\sqrt{E_s}\right)P\left(\sqrt{E_s}\right) + f\left(r_n|-\sqrt{E_s}\right)P\left(-\sqrt{E_s}\right) \tag{2.44}$$

where the probabilities $P\left(\sqrt{E_s}\right)$ and $P\left(-\sqrt{E_s}\right)$ correspond to $P(c_n = 0)$ and $P(c_n = 1)$, respectively. Assuming equal prior probabilities $P\left(c_n = 0\right) = P\left(c_n = 1\right) = \frac{1}{2}$, so that $P\left(\sqrt{E_s}\right) = P\left(-\sqrt{E_s}\right) = \frac{1}{2}$, it is easily shown (problem 7) by substituting Eqs. 2.42, 2.43 and 2.44 into Eq. 2.41 that

$$P\left(c_n = 0|r_n\right) = \frac{1}{1 + \exp\left[\frac{-2(\sqrt{E_s})r_n}{\sigma^2}\right]} \tag{2.45}$$

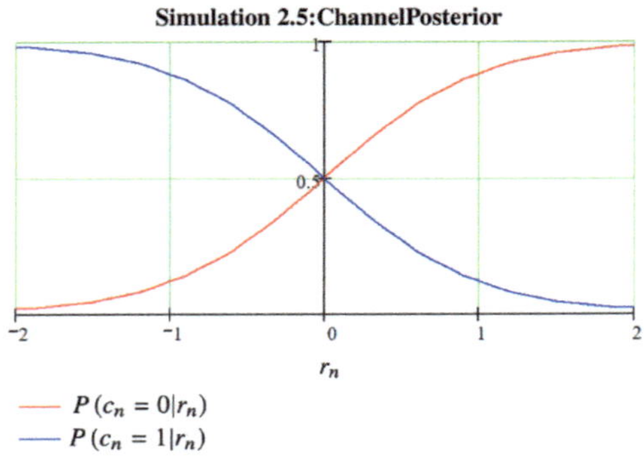

Fig. 2.12 Channel posterior probabilities

and similarly,

$$P(c_n = 1|r_n) = \frac{1}{1 + \exp\left[\frac{2(\sqrt{E_s})r_n}{\sigma^2}\right]} \tag{2.46}$$

These channel posterior probabilities are illustrated in Fig. 2.12 for $E_s = 1$ joule at a channel SNR = 0 dB.

As expected, for a large positive value of r_n, $P(c_n = 0|r_n)$ is a value close to 1 indicating a very high probability that the transmitted code digit c_n was a binary digit 0. If r_n is a large negative value, then $P(c_n = 0|r_n)$ is almost 0 because its very likely that the transmitted code digit c_n was a binary digit 1. If $r_n = 0$, there is no certainty as to whether c_n is 1 or a 0 and therefore as expected, $P(c_n = 0|r_n = 0) = 0.5$. Having found expressions for the channel posterior probabilities, substituting Eqs. 2.45 and 2.46 into Eq. 2.34 and simplifying (problem 8) gives the belief

$$\lambda_n = L_c r_n \tag{2.47}$$

where L_c is referred to as the *channel reliability* defined by

$$L_c = \frac{2\sqrt{E_s}}{\sigma^2} \tag{2.48}$$

The variation of λ_n is presented in Fig. 2.13 for $SNR = 0\,\text{dB}$ with $R_{code} = 1/2$, $E_b = 2$ joules for which $L_c = 2$. A positive value of λ_n suggests that a binary digit 0 was transmitted, a negative value is evidence in favor of a binary digit 1 being transmitted and the magnitude $|\lambda_n|$ is a measure of confidence.

Substituting $\lambda_n = L_c r_n$ into Eq. 2.33, the belief function $f(\lambda_1, \lambda_2)$ for the left child is given by $2 \cdot \tanh^{-1}\left(\tanh\left(\frac{L_c r_1}{2}\right) \cdot \tanh\left(\frac{L_c r_2}{2}\right)\right)$, where $L_c r_1$ and $L_c r_2$ are beliefs or equivalently the *extrinsic* information on c_1 and c_2 as we shall discover in the next chapter. For a given channel, L_c is a scaling factor. Therefore, in practice, only the value of r_n is used as the belief at the top level of the tree. In general, for a

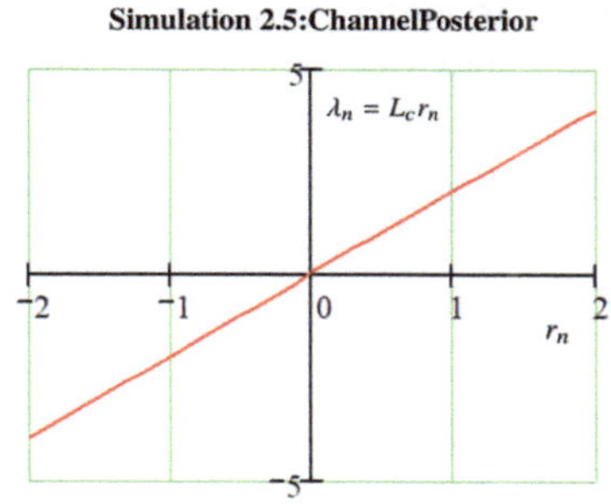

Fig. 2.13 λ_n variation for $E_s = 1$ joule. Simulation ChannelPosterior

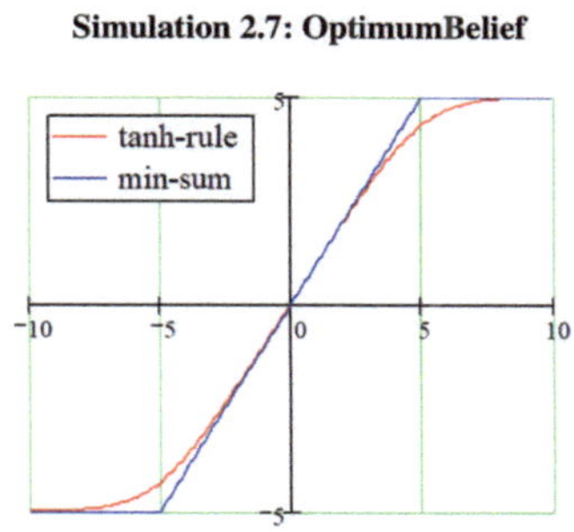

Fig. 2.14 Tanh-rule and min-sum comparison

given parent node with beliefs a and b within the polar decoder, the optimal belief for the left child is given by

$$f(a, b) = 2 \cdot \tanh^{-1}\left(\tanh\left(\frac{a}{2}\right) \cdot \tanh\left(\frac{b}{2}\right)\right) \tag{2.49}$$

A comparison of the belief function $f(a, b) = sign(a)sign(b)min(|a|, |b|)$, referred to as the *min-sum* approximation, versus the tanh rule of Eq. 2.49 for $b = 5$ as a ranges from -10 to 10 along the horizontal axis is presented in Fig. 2.14. Over the central region, the difference is insignificant, and this is also reflected in the BER performance comparison shown in Fig. 2.15. Clearly, there is no significant difference between the two BER curves. Therefore, the min-sum rule is used in practice because of its complexity efficiency with near-optimal performance.

2.8.2 Simulation: AWGN Channel Posterior Probability

In simulation channel posterior, Eq. 2.45 for the probability $P(c_n = 0|r_n)$ is verified by transmitting code digits with probability $P(c_n = 1) = P(c_n = 0) = \frac{1}{2}$ using BPSK over an AWGN channel. The noiseless BPSK signal point coordinates $+\sqrt{E_s}$ and $-\sqrt{E_s}$ correspond to the transmission of binary code digits $c_n = 0$ and $c_n = 1$, respectively. However, with noise, let x_h represent the coordinate of the received

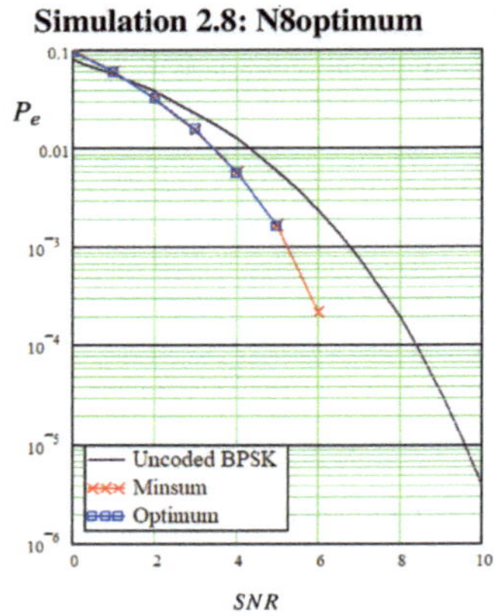

Fig. 2.15 BER Comparison of $f(.)$ functions

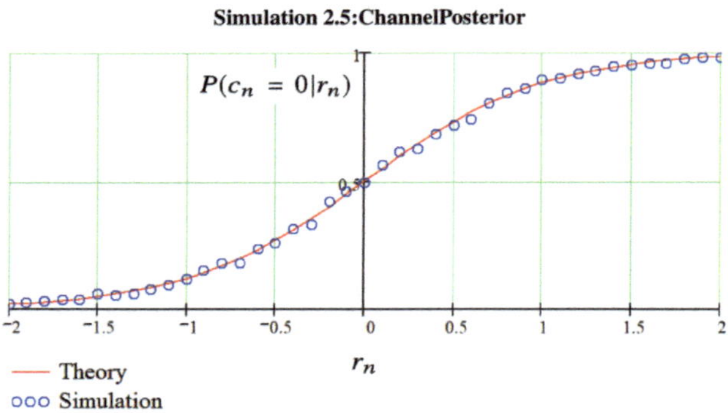

Fig. 2.16 Channel posterior probability $P(c_n = 0|r_n)$ for BPSK over an AWGN channel with $\Delta = 0.05$, $E_b = 2$ joule and $R_{code} = 0.5$

signal point, i.e., $x_h = \pm\sqrt{E_s} + n_n$. A counter that records the number of trials N_{trials} and a counter that records the number of events N_{events} are initialized to 0.

For a given value of r_n, if $(r_n - \Delta) \leq x_h \leq (r_n + \Delta)$, where Δ is a small number (e.g., 0.05), then N_{trials} is incremented, and if $c_n = 0$, then N_{events} is incremented. Finally, the simulation estimate of $P(c_n = 0|r_n)$ is given by $\left(\frac{N_{events}}{N_{trials}}\right)$. Figure 2.16 shows the results of this simulation for r_n ranging from -2 to $+2$ in steps of 0.1, overlaid with the theoretical variation given by Eq. 2.45. As expected, we have excellent agreement between simulation and theory, provided N_{trials} is sufficiently large to ensure an accurate estimate.

2.9 Appendix: Tanh Rule

In Sect. 2.8, the equivalence of Eqs. 2.32 and 2.33 that was stated is verified in this appendix. From the definition of the hyperbolic tangent

$$\tanh(u) = \frac{e^u - e^{-u}}{e^u + e^{-u}} \tag{2.50}$$

Rearranging (problem 9)

$$e^{2u} = \frac{1 + \tanh(u)}{1 - \tanh(u)} \tag{2.51}$$

For $2u = \lambda_1$

$$e^{\lambda_1} = \frac{1 + \tanh\left(\frac{\lambda_1}{2}\right)}{1 - \tanh\left(\frac{\lambda_1}{2}\right)} = \frac{1 + x}{1 - x} \tag{2.52}$$

where $x = \tanh\left(\frac{\lambda_1}{2}\right)$. Similarly, for $y = \tanh\left(\frac{\lambda_2}{2}\right)$

$$e^{\lambda_2} = \frac{1 + \tanh\left(\frac{\lambda_2}{2}\right)}{1 - \tanh\left(\frac{\lambda_2}{2}\right)} = \frac{1 + y}{1 - y} \tag{2.53}$$

Substituting these into Eq. 2.32 for $\lambda(c_1 \oplus c_2)$

$$\lambda(c_1 \oplus c_2) = \ln\left(\frac{\left(\frac{1+x}{1-x}\right)\left(\frac{1+y}{1-y}\right) + 1}{\frac{1+x}{1-x} + \frac{1+y}{1-y}}\right) \tag{2.54}$$

Simplifying the numerator and denominator separately

$$\text{Numerator: } \frac{(1 + x)(1 + y)}{(1 - x)(1 - y)} + 1 = \frac{2(1 + xy)}{(1 - x)(1 - y)} \tag{2.55}$$

$$\text{Denominator: } \frac{1 + x}{1 - x} + \frac{1 + y}{1 - y} = \frac{2(1 - xy)}{(1 - x)(1 - y)} \tag{2.56}$$

Combining these results

$$\lambda(c_1 \oplus c_2) = \ln\left(\frac{1 + xy}{1 - xy}\right) \tag{2.57}$$

Now using the inverse hyperbolic tangent function

$$\tanh^{-1}(u) = \frac{1}{2}\ln\left(\frac{1+u}{1-u}\right) \tag{2.58}$$

For $u = xy$

$$\lambda(c_1 \oplus c_2) = \ln\left(\frac{1+xy}{1-xy}\right) = 2 \cdot \tanh^{-1}(xy) \tag{2.59}$$

and replacing x and y

$$\lambda(c_1 \oplus c_2) = \ln\left(\frac{P(c_1 \oplus c_2 = 0)}{P(c_1 \oplus c_2 = 1)}\right) = 2 \cdot \tanh^{-1}\left(\tanh\left(\frac{\lambda_1}{2}\right) \cdot \tanh\left(\frac{\lambda_2}{2}\right)\right) \tag{2.60}$$

where $\lambda_1 = \ln\left(\frac{P(c_1=0)}{P(c_1=1)}\right)$ and $\lambda_2 = \ln\left(\frac{P(c_2=0)}{P(c_2=1)}\right)$.

2.10 Appendix: Successive Cancellation List (SCL) Decoding

Up to this point, we have considered only the Successive Cancellation (SC) decoding algorithm in which the binary digits u_1 to u_N are decided one after another along a single path through the decoding tree. A drawback of SC decoding is that if a wrong decision is made at any stage, then all subsequent belief updates are computed using an incorrect value, which leads to error propagation. To minimize this effect, *Successive Cancellation List* (SCL) decoding is used in practice. Instead of maintaining a single path, the SCL decoder tracks L concurrent decoding paths. At each information bit, the decoder branches and considers both possible decisions 0 and 1. To prevent the number of paths from growing to 2^N, the decoder sorts the paths at each step according to a reliability metric and keeps only the L most likely paths.

2.10.1 Path Metric Update

Each path p is assigned a path metric $\mathrm{PM}^{(p)}$, which accumulates a penalty whenever a decision contradicts the sign of the corresponding belief. At the start of decoding

$$\mathrm{PM}^{(p)} = 0 \tag{2.61}$$

for every path. Let L_i denote the belief (LLR) calculated for bit u_i on a given path. The sign of L_i indicates the preferred decision with positive values favoring 0 and negative values favoring 1. The magnitude $|L_i|$ measures the reliability of that preference. For simplicity, we shall use the following approximation to update the path metric. When a decision $u_i^{(p)}$ is taken on path p, the path metric is updated as

$$\mathrm{PM}^{(p)} \leftarrow \mathrm{PM}^{(p)} + \Delta\mathrm{PM}(u_i^{(p)}) \tag{2.62}$$

where

$$\Delta\mathrm{PM}(u_i^{(p)}) = \begin{cases} 0 & \text{if } \mathit{Decision}(L_i) = u_i^{(p)} \\ |L_i| & \text{if } \mathit{Decision}(L_i) \neq u_i^{(p)} \end{cases} \tag{2.63}$$

A path that agrees with the sign of the belief receives no penalty. A path that contradicts that the sign receives a penalty equal to the magnitude of the belief. Large-magnitude beliefs therefore strongly discourage the less likely decision and paths with smaller path metrics are more likely to be correct. A more accurate path metric update is

$$\Delta\mathrm{PM}(u_i^{(p)}) = \ln\Big(1 + \exp\big(-(1 - 2u_i^{(p)})L_i\big)\Big) \tag{2.64}$$

2.10.2 *Decoding Walk-through:* (8, 4) *Code with* $L = 2$

Consider the standard SCL algorithm using the (8, 4) polar code described earlier. The frozen set is

$$\mathcal{F} = \{1, 2, 3, 5\}$$

and the information set is

$$\mathcal{I} = \{4, 6, 7, 8\}$$

so that

$$u_1 = u_2 = u_3 = u_5 = 0, \quad u_4 = m_1, \quad u_6 = m_2, \quad u_7 = m_3, \quad u_8 = m_4$$

The decoder uses the same belief matrix and the same $f(\cdot)$ and $g(\cdot)$ functions as in the SC description in Sect. 2.5. The only difference is that several paths are retained in parallel. We focus only on the decisions at the four information positions and assume that all paths see the same beliefs at these positions

$$L_4 = 1.20, \quad L_6 = -1.50, \quad L_7 = 0.30, \quad L_8 = -2.00 \tag{2.65}$$

Positive values favor binary digit 0, negative values favor binary digit 1, and the magnitude indicates the strength of the preference. The list size is $L = 2$. A compact graphical view of the same example is given in the flowchart in Fig. 2.17, which shows how the active paths and their metrics evolve across the information bits.

Frozen Bits u_1, u_2, u_3, u_5

At the frozen positions $i \in \mathcal{F}$, there is no branching. Every active path is extended with the forced decision $u_i^{(p)} = 0$ and the beliefs in the tree are updated using the same SC rules. However, the path metric should still be updated at frozen indices because the forced decision may contradict the sign of the corresponding belief. Using the same penalty rule as for information bits, the update at a frozen position is

$$\mathrm{PM}^{(p)} \leftarrow \mathrm{PM}^{(p)} + \Delta\mathrm{PM}\big(u_i^{(p)} = 0\big)$$

where

$$\Delta\mathrm{PM}\big(u_i^{(p)} = 0\big) = \begin{cases} 0 & \text{if } \mathit{Decision}(L_i^{(p)}) = 0 \\ |L_i^{(p)}| & \text{if } \mathit{Decision}(L_i^{(p)}) = 1 \end{cases}$$

This ensures that paths are penalized whenever the forced frozen-bit decision disagrees with the sign of the belief. For simplicity, we shall assume that the beliefs at the information positions are the same across paths. In general, after paths diverge, each path p has its own beliefs $L_i^{(p)}$ because the $g(\cdot)$ updates depend on previous decisions.

Step 1: Bit u_4 (Information)

At u_4, the single existing path splits into two paths:

- Path P_1: $u_4^{(1)} = 0$. This agrees with $L_4 > 0$ so $\Delta\mathrm{PM} = 0$ and the path metric remains 0.
- Path P_2: $u_4^{(2)} = 1$. This contradicts the sign of L_4 so $\Delta\mathrm{PM} = |L_4| = 1.20$ and the path metric becomes 1.20.

Both paths are kept because the list size is $L = 2$.

Step 2: Bit u_6 (Information)

After processing the frozen bit u_5, each of the two paths reaches bit u_6 with the same belief $L_6 = -1.50$, which favors $u_6 = 1$. Each path now splits into two paths.

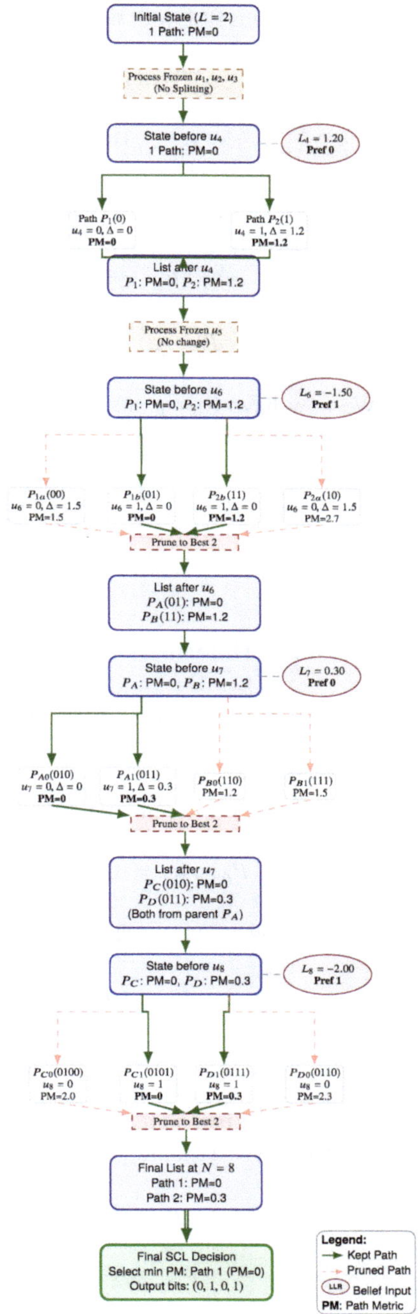

Fig. 2.17 Flow chart of the SCL decoding process with $L = 2$ for the (8, 4) example

From P_1 with PM $= 0$, we obtain

$$P_{1a} : u_6^{(1a)} = 0, \quad \Delta\text{PM} = |L_6| = 1.50, \quad \text{PM}^{(1a)} = 1.50$$

$$P_{1b} : u_6^{(1b)} = 1, \quad \Delta\text{PM} = 0, \quad \text{PM}^{(1b)} = 0$$

From P_2 with PM $= 1.20$, we obtain

$$P_{2a} : u_6^{(2a)} = 0, \quad \Delta\text{PM} = 1.50, \quad \text{PM}^{(2a)} = 2.70$$

$$P_{2b} : u_6^{(2b)} = 1, \quad \Delta\text{PM} = 0, \quad \text{PM}^{(2b)} = 1.20$$

We now have four paths. The two with the smallest metrics are P_{1b} and P_{2b}. After pruning, the list contains

$$P_A : (u_4, u_6) = (0, 1), \quad \text{PM} = 0$$

$$P_B : (u_4, u_6) = (1, 1), \quad \text{PM} = 1.20$$

Step 3: Bit u_7 (Information)

At u_7, both paths see the same belief $L_7 = 0.30$, which favors $u_7 = 0$.

From P_A with PM $= 0$

$$P_{A0} : u_7 = 0, \quad \Delta\text{PM} = 0, \quad \text{PM} = 0$$

$$P_{A1} : u_7 = 1, \quad \Delta\text{PM} = |L_7| = 0.30, \quad \text{PM} = 0.30$$

From P_B with PM $= 1.20$

$$P_{B0} : u_7 = 0, \quad \Delta\text{PM} = 0, \quad \text{PM} = 1.20$$

$$P_{B1} : u_7 = 1, \quad \Delta\text{PM} = 0.30, \quad \text{PM} = 1.50$$

The two best paths are P_{A0} and P_{A1} so both surviving paths descend from P_A. After pruning, we rename

$$P_C : (u_4, u_6, u_7) = (0, 1, 0), \quad \text{PM} = 0$$

$$P_D : (u_4, u_6, u_7) = (0, 1, 1), \quad \text{PM} = 0.30$$

Step 4: Bit u_8 (Information)

Finally, at u_8, the belief $L_8 = -2.00$ favors $u_8 = 1$.
From P_C with PM $= 0$

$$P_{C0} : u_8 = 0, \quad \Delta\text{PM} = |L_8| = 2.00, \quad \text{PM} = 2.00$$

$$P_{C1} : u_8 = 1, \quad \Delta\text{PM} = 0, \quad \text{PM} = 0$$

From P_D with PM $= 0.30$

$$P_{D0} : u_8 = 0, \quad \Delta\text{PM} = 2.00, \quad \text{PM} = 2.30$$

$$P_{D1} : u_8 = 1, \quad \Delta\text{PM} = 0, \quad \text{PM} = 0.30$$

At the end of decoding the two best paths are P_{C1} and P_{D1} with metrics 0 and 0.30. The SCL decoder now selects P_{C1} because it has the smallest path metric. The corresponding information bits are

$$(\hat{u}_4, \hat{u}_6, \hat{u}_7, \hat{u}_8) = (0, 1, 0, 1)$$

so that

$$[\hat{m}_1, \hat{m}_2, \hat{m}_3, \hat{m}_4] = [0, 1, 0, 1]$$

This example and the flowchart in Fig. 2.17 together illustrate how the SCL decoder maintains a small list of paths through the polar decoding tree and uses a simple path metric to recover from early wrong decisions.

2.10.3 CRC-Aided SCL (CA-SCL)

While SCL decoding provides an improved performance over SC, the most likely path according to the path metric is not always the correct one. A simple and effective refinement is to concatenate a *Cyclic Redundancy Check* (CRC) with the polar code. The CRC is an error-detecting code that treats the binary data as a polynomial. The message $M(x)$ is shifted by the degree r of a generator polynomial $G(x)$ and then divided by $G(x)$ in GF(2). The remainder $R(x)$ is appended to the message to form the transmitted polynomial

$$T(x) = M(x)x^r + R(x)$$

At the receiver, any valid codeword must be exactly divisible by $G(x)$. For example, consider a message $M = 110101$ and generator $G = 1011$ of degree $r = 3$.

Table 2.5 BER comparison: uncoded BPSK and polar decoding schemes

E_b/N_0 (dB)	Uncoded	SC	SCL	SCL + CRC
0.0	7.86e-02	9.55e-02	9.08e-02	5.93e-02
1.0	5.63e-02	6.15e-02	5.90e-02	3.70e-02
2.0	3.75e-02	3.18e-02	2.95e-02	1.65e-02
3.0	2.29e-02	2.03e-02	1.76e-02	1.24e-02
4.0	1.25e-02	5.86e-03	5.65e-03	2.62e-03
5.0	5.95e-03	1.35e-03	1.13e-03	5.50e-04
6.0	2.39e-03	2.90e-04	2.00e-04	1.20e-04

- Append three 0s to form 110101000.
- Divide 110101000 by 1011 using modulo-2 division. The remainder is 111.
- The transmitted CRC-protected word is 110101111.

In CA-SCL decoding, the C CRC bits are appended to the information bits before polar encoding. The SCL decoder then produces up to L path estimates of the information word at the output. Instead of always selecting the path with the lowest path metric, the decoder first discards any paths whose reconstructed information word fails the CRC check. Among the remaining CRC-valid paths, the one with the smallest path metric is selected as the final output. If no path passes the CRC, then the decoder may fall back to the path with the smallest metric or declare a failure depending on the system design.

2.10.4 *Simulation* (8, 4) *Polar Code SCL Decoding with CRC*

The bit error rate (BER) as a function of E_b/N_0 is summarized in Table 2.5 using BPSK over an AWGN channel comparing SC, SCL with list size $L = 8$, and CRC-aided SCL (CA-SCL).

Notice the progressive performance gains achieved by SC, SCL, and CA-SCL decoding. As the SNR increases, the list decoder with $L = 8$ begins to outperform SC because the correct path is more often retained in the list. The addition of CRC bits introduces a fixed overhead, which is negligible for long block lengths but may be significant for short control blocks so there is a trade-off between throughput and reliability.

2.11 Appendix: 5G Standard Reliability Sequence for $N \leq 1024$

1, 2, 3, 5, 9, 17, 33, 4, 6, 65, 10, 7, 18, 11, 19, 129, 13, 34, 66, 21, 257, 35, 25, 37, 8, 130, 67, 513, 12, 41, 69, 131, 20, 14, 49, 15, 73, 258, 22, 133, 36, 259, 27, 514, 81, 38, 26, 23, 137, 261, 265, 39, 515, 97, 68, 42, 145, 29, 70, 43, 517, 50, 75, 273, 161,

521, 289, 529, 193, 545, 71, 45, 132, 82, 51, 74, 16, 321, 134, 53, 24, 135, 385, 77, 138, 83, 57, 28, 98, 40, 260, 85, 139, 146, 262, 30, 44, 99, 516, 89, 141, 31, 147, 72, 263, 266, 162, 577, 46, 101, 641, 52, 149, 47, 76, 267, 274, 518, 105, 163, 54, 194, 153, 78, 165, 769, 269, 275, 519, 55, 84, 58, 522, 113, 136, 79, 290, 195, 86, 277, 523, 59, 169, 140, 100, 87, 61, 281, 90, 291, 530, 525, 197, 142, 102, 148, 177, 143, 531, 322, 32, 201, 91, 546, 293, 323, 533, 264, 150, 103, 106, 305, 297, 164, 93, 48, 268, 386, 547, 325, 209, 387, 151, 154, 166, 107, 56, 329, 537, 578, 549, 114, 155, 80, 270, 109, 579, 225, 167, 520, 553, 196, 271, 642, 524, 276, 581, 292, 60, 170, 561, 115, 278, 157, 88, 198, 117, 171, 62, 532, 526, 643, 282, 279, 527, 178, 294, 389, 92, 585, 770, 199, 173, 121, 202, 337, 63, 283, 144, 104, 179, 295, 94, 645, 203, 593, 324, 393, 298, 771, 108, 181, 152, 210, 285, 649, 95, 205, 299, 401, 609, 353, 326, 534, 156, 211, 306, 548, 301, 110, 185, 535, 538, 116, 168, 226, 327, 307, 773, 158, 657, 330, 111, 118, 213, 172, 777, 331, 227, 550, 539, 388, 309, 217, 417, 272, 280, 159, 338, 551, 673, 119, 333, 580, 541, 390, 174, 122, 554, 200, 785, 180, 229, 339, 313, 705, 391, 175, 555, 582, 394, 284, 123, 449, 354, 562, 204, 64, 341, 395, 528, 583, 557, 182, 296, 286, 233, 125, 206, 183, 644, 563, 287, 586, 300, 355, 212, 402, 186, 397, 345, 587, 646, 594, 536, 241, 207, 96, 328, 565, 801, 403, 357, 308, 302, 418, 214, 569, 833, 589, 187, 647, 405, 228, 897, 595, 419, 303, 650, 772, 361, 540, 112, 332, 215, 310, 189, 450, 218, 409, 610, 597, 552, 651, 230, 160, 421, 311, 542, 774, 611, 658, 334, 120, 601, 340, 219, 369, 653, 231, 392, 314, 451, 543, 335, 234, 556, 775, 176, 124, 659, 613, 342, 778, 221, 315, 425, 396, 674, 584, 356, 288, 184, 235, 126, 558, 661, 617, 343, 317, 242, 779, 564, 346, 453, 398, 404, 208, 675, 559, 786, 433, 358, 188, 237, 665, 625, 588, 781, 706, 127, 243, 566, 399, 347, 457, 359, 406, 304, 570, 245, 596, 190, 567, 677, 362, 707, 590, 216, 787, 648, 349, 420, 407, 465, 681, 802, 363, 591, 410, 571, 789, 598, 573, 220, 312, 709, 599, 602, 652, 422, 793, 803, 612, 603, 411, 232, 689, 654, 249, 370, 191, 365, 655, 660, 336, 481, 316, 222, 371, 614, 423, 426, 452, 615, 544, 236, 413, 344, 373, 776, 318, 223, 427, 454, 238, 560, 834, 805, 713, 835, 662, 809, 780, 618, 605, 434, 721, 817, 837, 348, 898, 244, 663, 455, 319, 676, 619, 899, 782, 377, 429, 666, 737, 568, 841, 626, 239, 360, 458, 400, 788, 592, 679, 435, 678, 350, 246, 459, 667, 621, 364, 128, 192, 783, 408, 437, 627, 572, 466, 682, 247, 708, 351, 600, 669, 791, 461, 250, 683, 574, 412, 804, 790, 710, 366, 441, 629, 690, 375, 424, 467, 794, 251, 372, 482, 575, 414, 604, 367, 469, 656, 901, 806, 616, 685, 711, 430, 795, 253, 374, 606, 849, 691, 714, 633, 483, 807, 428, 905, 415, 224, 664, 693, 836, 620, 473, 456, 797, 810, 715, 722, 838, 717, 865, 811, 607, 913, 723, 697, 378, 436, 818, 320, 622, 813, 485, 431, 839, 668, 489, 240, 379, 460, 623, 628, 438, 381, 819, 462, 497, 670, 680, 725, 842, 630, 352, 468, 439, 738, 252, 463, 443, 442, 470, 248, 684, 843, 739, 900, 671, 784, 850, 821, 729, 929, 792, 368, 902, 631, 686, 845, 634, 712, 254, 692, 825, 903, 687, 741, 851, 376, 445, 471, 484, 416, 486, 906, 796, 474, 635, 745, 853, 961, 866, 694, 798, 907, 716, 808, 475, 637, 695, 255, 718, 576, 914, 799, 812, 380, 698, 432, 608, 490, 867, 724, 487, 909, 719, 814, 477, 857, 840, 726, 699, 915, 753, 869, 820, 815, 440, 930, 491, 624, 672, 740, 917, 464, 844, 382, 498, 931, 822, 727, 962, 873, 493, 632, 730, 701, 444, 742, 846, 921, 383, 823, 852, 731, 499, 881, 743, 446, 472, 636, 933, 688, 904, 826, 501, 847, 746, 827, 733, 447, 963, 937, 476, 854, 868, 638, 908, 488, 696, 747, 829, 754, 855, 858, 505, 800, 256, 965, 910, 720, 478, 916, 639, 749,

945, 870, 492, 700, 755, 859, 479, 969, 384, 911, 816, 977, 871, 918, 728, 494, 874, 702, 932, 757, 861, 500, 732, 824, 923, 875, 919, 503, 934, 744, 761, 882, 495, 703, 922, 502, 877, 848, 993, 448, 734, 828, 935, 883, 938, 964, 748, 506, 856, 925, 735, 830, 966, 939, 885, 507, 750, 946, 967, 756, 860, 941, 831, 912, 872, 640, 889, 480, 947, 751, 970, 509, 862, 758, 971, 920, 876, 863, 759, 949, 978, 924, 973, 762, 878, 953, 496, 704, 936, 979, 884, 763, 504, 926, 879, 736, 994, 886, 940, 995, 981, 927, 765, 942, 968, 887, 832, 948, 508, 890, 985, 752, 943, 997, 972, 891, 510, 950, 974, 1001, 893, 951, 864, 760, 1009, 511, 980, 954, 764, 975, 955, 880, 982, 983, 928, 996, 766, 957, 888, 986, 998, 987, 944, 892, 999, 767, 512, 989, 1002, 952, 1003, 894, 976, 895, 1010, 956, 1005, 1011, 958, 984, 959, 988, 1013, 1000, 1017, 768, 990, 1004, 991, 1006, 960, 1012, 1014, 896, 1007, 1015, 1018, 1019, 992, 1021, 1008, 1016, 1020, 1022, 1023, 1024.

2.12 Problems

1. Consider a polar (4,2) code, where the first two bits are frozen and the last two bits contain the message [0, 0]. Assuming transmission over a binary symmetric channel (BSC) with a crossover probability α, derive an expression for the probability of error on the first message bit.
2. Using the recursive formula for the erasure probabilities of synthetic channels after each polarization step, calculate the erasure probabilities for the four synthetic channels of a polar code with $N = 4$.
3. Determine all the Hamming weights of all the codewords of the rate 1/2 (8, 4) polar code. Comment on your result.
4. Suppose an (8, 4) polar code is transmitted over an AWGN channel with an SNR of 3 dB. The received LLRs are $[1.2, -0.8, 0.6, -1.4, 1.5, -1.0, 0.4, -1.2]$. Perform successive cancellation decoding for this received word and determine the decoded message bits. Explain each step of the decoding process and highlight the impact of frozen bits.
5. Each split in the polar decoding tree represents the creation of two synthetic channels. Assume that the original channel is a binary erasure channel (BEC) with erasure probability p. In the polar transform, the synthetic channels can be modeled as two new BECs with different erasure probabilities. Show that the erasure probabilities of the left and right synthetic channels are $2p - p^2$ and p^2, respectively.
6. Show that $P(c_n = 0|r_n) = \frac{P(r_n|c_n=0)P(c_n=0)}{P(r_n)}$ is equivalent to $P(c_n = 0|r_n) = \frac{f(r_n|\sqrt{E_s})P(c_n=0)}{f(r_n)}$, where $f\left(r_n|\sqrt{E_s}\right) = \frac{1}{\sigma\sqrt{2\pi}} \exp\left[\frac{-(r_n-\sqrt{E_s})^2}{2\sigma^2}\right]$.

7. Substitute Eqs. 2.42, 2.43 and 2.44 into Eq. 2.41 to show that $P(c_n = 0|r_n) = \frac{1}{1+\exp\left[\frac{-2(\sqrt{E_s})r_n}{\sigma^2}\right]}$ and $P(c_n = 1|r_n) = \frac{1}{1+\exp\left[\frac{2(\sqrt{E_s})r_n}{\sigma^2}\right]}$.
8. Substitute Eq. 2.45 and 2.46 into Eq. 2.34 to show that $\lambda_n = L_c r_n$.
9. Given $\tanh(u) = \frac{e^u - e^{-u}}{e^u + e^{-u}}$, show that $e^{2u} = \frac{1+\tanh(u)}{1-\tanh(u)}$.
10. For a (16, 8) polar code, construct the polar code tree and derive the reliability sequence using the Bhattacharyya parameters. Compute the synthetic channel reliabilities for each stage and list the positions for the frozen and message bits, similar to the (8, 4) case in the chapter. Compare the results with the reliability sequence for the 5G standard for $N = 16$.
11. If a polar code with a block length $N = 32$ and rate $\frac{1}{4}$ is constructed, which indices should be used for the information bits using the 5G standard?
12. Show that $\lambda(c_1 \oplus c_2) = -2 \cdot \tanh^{-1}\left(\tanh\left(\frac{-\lambda_1}{2}\right) \cdot \tanh\left(\frac{-\lambda_2}{2}\right)\right)$ if $\lambda_1 = \ln\left(\frac{P(c_1=1)}{P(c_1=0)}\right)$ and $\lambda_2 = \ln\left(\frac{P(c_2=1)}{P(c_2=0)}\right)$. Note the reversal of the LLR in this problem.
13. Using the hyperbolic tangent formula for $\lambda(c_1 \oplus c_2)$, derive the expression for $\lambda(c_1 \oplus c_2 \oplus c_3)$ for both the standard and reversed definitions of an LLR.
14. Consider a binary-input additive white Gaussian noise (AWGN) channel with BPSK modulation. A transmitted bit $c \in \{0, 1\}$ is mapped to a signal $x \in \{+\sqrt{E_s}, -\sqrt{E_s}\}$. The received value is $r = x + w$, where $w \sim \mathcal{N}(0, \sigma^2)$.

    (a) Show that the log-likelihood ratio (LLR) λ for the bit c given the received value r satisfies

    $$\lambda = \ln\left(\frac{P(c = 0 \mid r)}{P(c = 1 \mid r)}\right) = \frac{2r\sqrt{E_s}}{\sigma^2}$$

    Assume equal prior probabilities $P(c = 0) = P(c = 1) = \frac{1}{2}$.

    (b) Explain why, in a practical implementation of a successive cancellation (SC) polar decoder, the beliefs $L_{1,j} = r_j$ are used directly for $j = 1, \ldots, N$, where r_j is the received value corresponding to the j-th transmitted code bit.

15. In a successive cancellation (SC) decoder for a polar code, consider a node with the code digits

    $$c_1 = u_a \oplus u_b \qquad c_2 = u_b$$

    Assume that the bit u_a has already been decoded, and its decision is denoted by $\hat{u}_a = i \in \{0, 1\}$. Let r_1 and r_2 be the received values corresponding to c_1 and c_2, respectively. Given the log-likelihood ratios (LLRs) of these bits,

    $$a = \lambda(c_1) = \ln\left(\frac{P(c_1 = 0 \mid r_1)}{P(c_1 = 1 \mid r_1)}\right) \qquad b = \lambda(c_2) = \ln\left(\frac{P(c_2 = 0 \mid r_2)}{P(c_2 = 1 \mid r_2)}\right)$$

show that the conditional LLR used to decode u_b, given the prior decision $\hat{u}_a = i$, is

$$g(a, b, i) = \ln\left(\frac{P(u_b = 0 \mid r_1, r_2, \hat{u}_a = i)}{P(u_b = 1 \mid r_1, r_2, \hat{u}_a = i)}\right) = b + (1 - 2i)a$$

Chapter 3
Low-Density Parity Check (LDPC) Codes

Developed by Robert Gallager in 1960 as part of his Ph.D. thesis at the Massachusetts Institute of Technology (MIT), *low-density parity check* (LDPC) codes or *Gallager* codes are a class of linear block codes. These codes were largely ignored because they were impractical to implement in those days. With the advent of Turbo codes in 1993, which demonstrated the use of iterative decoding to achieve a performance close to that predicted by Shannon, a renewed interest in iterative decoding techniques brought LDPC codes into the forefront of error-control coding research. For example, Richardson et al. [9] presented results for BPSK over an AWGN channel that operates very close to Shannon's capacity prediction. Surpassing turbo codes for numerous applications and standards, LDPC codes are now a part of widely used standards including IEEE 802.11 (Wi-Fi), IEEE 802.3 (10GBASE-T Ethernet), and DVB-S2. However, an LDPC decoder requires many more iterations than a turbo decoder, which is a disadvantage for latency critical applications. LDPC codes require the specification of the parity-check matrix $\mathbf{H}$ from which the generator matrix $\mathbf{G}$ is derived. Although a number of decoding algorithms have been proposed for LDPC codes, we will only consider the most popular decoding algorithm for *binary* LDPC codes.

We begin with the fundamentals of block codes in Sect. 3.1 by considering the (7, 4) Hamming code, invented in 1950 by Richard W. Hamming, which led to the development of other linear block codes such as Reed–Solomon codes, BCH codes, and Golay codes. Then, a graphical representation of a simple parity-check matrix $\mathbf{H}$ is presented in Sect. 3.2, which forms the foundation of LDPC codes. The LDPC decoding algorithm is presented in Sect. 3.3, and a simple LDPC code is introduced in Sect. 3.4, where the encoding and decoding steps are illustrated in full detail. Finally, in Sect. 3.5, WiMAX LDPC codes are discussed, including a general encoder and the simulation performance of the WiMAX 802.16e rate 1/2 (288×576) LDPC code.

J. Sodha, *Fundamentals of Polar, LDPC and Turbo Codes*,
https://doi.org/10.1007/978-3-031-90924-5_3

3.1 Fundamentals of Linear Block Codes

For a general (N, K) block code, a codeword $\mathbf{c}$ is given by

$$\mathbf{c} = \mathbf{m}\mathbf{G} \tag{3.1}$$

where $\mathbf{m}$ is the message (or information) vector and $\mathbf{G}$ is the generator matrix with N columns and K rows. In a *systematic* block code, the generator matrix $\mathbf{G}$ consists of a parity matrix $\mathbf{P}$ appended to an identity matrix $\mathbf{I}_K$ such that

$$\mathbf{G} = [\mathbf{I}_K\ \mathbf{P}] \tag{3.2}$$

For example, consider the generator matrix of a (7, 4) given by

$$\mathbf{G} = \begin{bmatrix} 1\,0\,0\,0\,1\,1\,0 \\ 0\,1\,0\,0\,1\,0\,1 \\ 0\,0\,1\,0\,0\,1\,1 \\ 0\,0\,0\,1\,1\,1\,1 \end{bmatrix} \tag{3.3}$$

where

$$\mathbf{I}_4 = \begin{bmatrix} 1\,0\,0\,0 \\ 0\,1\,0\,0 \\ 0\,0\,1\,0 \\ 0\,0\,0\,1 \end{bmatrix}, \quad \mathbf{P} = \begin{bmatrix} 1\,1\,0 \\ 1\,0\,1 \\ 0\,1\,1 \\ 1\,1\,1 \end{bmatrix} \tag{3.4}$$

For $\mathbf{m} = [1\ 0\ 1\ 0]$, the corresponding codeword is

$$\mathbf{c} = [1\ 0\ 1\ 0] \begin{bmatrix} 1\,0\,0\,0\,1\,1\,0 \\ 0\,1\,0\,0\,1\,0\,1 \\ 0\,0\,1\,0\,0\,1\,1 \\ 0\,0\,0\,1\,1\,1\,1 \end{bmatrix} = [1\ 0\ 1\ 0\ 1\ 0\ 1] \tag{3.5}$$

Notice that the codeword contains the original message as the first part. Since the input binary sequence can only be one of 2^K possible message words, only 2^K out of 2^N possible words are used as codewords.

A *linear* block code has the property that the modulo-2 sum of any two codewords produces another codeword. The *Hamming distance* between two binary words of the same length is the number of differing digits. For example, the Hamming distance between $\mathbf{c_1} = 1010101$ and $\mathbf{c_2} = 1110000$ is 3. The total number of ones in a codeword is called its *Hamming weight*. The distance between codewords and the error detection and correction capabilities of linear block codes are explored in the problems section. The syndrome **s** of a block code is defined by

$$\mathbf{s} = \mathbf{r}\mathbf{H}^T \tag{3.6}$$

where **r** is the received word, and **H** is a *parity-check matrix* such that

$$\mathbf{G}\mathbf{H}^T = \mathbf{0} \tag{3.7}$$

Let **e** represent the error vector, such that

$$\mathbf{r} = \mathbf{c} + \mathbf{e} \tag{3.8}$$

where **c** is the transmitted codeword. For example, if $\mathbf{c} = 1010101$ and $\mathbf{e} = 0000010$, then $\mathbf{r} = 1010111$. In this case,

$$\mathbf{s} = \mathbf{r}\mathbf{H}^T = (\mathbf{c} + \mathbf{e})\,\mathbf{H}^T = \mathbf{c}\mathbf{H}^T + \mathbf{e}\mathbf{H}^T \tag{3.9}$$

Since $\mathbf{c} = \mathbf{m}\mathbf{G}$, we find

$$\mathbf{s} = \mathbf{m}\mathbf{G}\mathbf{H}^T + \mathbf{e}\mathbf{H}^T = \mathbf{e}\mathbf{H}^T \tag{3.10}$$

For example, Table 3.1 presents the syndromes for single-digit errors in the received word (Problem 3). In this case, the syndromes are unique and can be used to correct the error within the received word **r**. However, for more than one error within the received word, the syndromes are not unique. Having covered the fundamentals, we next explore how the parity-check matrix can be viewed as a Tanner graph, which is key to the LDPC decoding algorithm.

Table 3.1 (7,4) Decoding table

Simulation 3.1: Syndrome74

Syndrome **s**			Error vector **e**						
0	0	0	0	0	0	0	0	0	0
0	0	1	0	0	0	0	0	0	1
0	1	0	0	0	0	0	0	1	0
1	0	0	0	0	0	0	1	0	0
1	1	1	0	0	0	1	0	0	0
0	1	1	0	0	1	0	0	0	0
1	0	1	0	1	0	0	0	0	0
1	1	0	1	0	0	0	0	0	0

3.2 Tanner Graph

Consider the parity check matrix **H** of size (5×10) given by

$$\mathbf{H} = \begin{bmatrix} 1\,1\,1\,0\,0\,1\,1\,0\,0\,1 \\ 1\,0\,1\,0\,1\,1\,0\,1\,1\,0 \\ 0\,0\,1\,1\,1\,0\,1\,0\,1\,1 \\ 0\,1\,0\,1\,1\,1\,0\,1\,0\,1 \\ 1\,1\,0\,1\,0\,0\,1\,1\,1\,0 \end{bmatrix} \tag{3.11}$$

Let $\mathbf{c} = \{c_0, c_1, \cdots, c_n, \cdots, c_{N-1}\}$ represent the transmitted codeword, and let $\mathbf{r} = \{r_0, r_1, \cdots, r_n, \cdots, r_{N-1}\}$ represent the corresponding received noisy word, where N is the length of the codeword or equivalently, the number of columns in **H**. In this example, the syndrome $\mathbf{s} = \mathbf{c}\mathbf{H}^T$ is given by

$$\mathbf{s} = \mathbf{c}\mathbf{H}^T = \begin{bmatrix} c_0 \; c_1 \; c_2 \; c_3 \; c_4 \; c_5 \; c_6 \; c_7 \; c_8 \; c_9 \end{bmatrix} \begin{bmatrix} 1\,1\,0\,0\,1 \\ 1\,0\,0\,1\,1 \\ 1\,1\,1\,0\,0 \\ 0\,0\,1\,1\,1 \\ 0\,1\,1\,1\,0 \\ 1\,1\,0\,1\,0 \\ 1\,0\,1\,0\,1 \\ 0\,1\,0\,1\,1 \\ 0\,1\,1\,0\,1 \\ 1\,0\,1\,1\,0 \end{bmatrix}$$

$$= \begin{bmatrix} s_0 \; s_1 \; s_2 \; s_3 \; s_4 \end{bmatrix} \tag{3.12}$$

where with modulo-2 addition

$$s_0 = c_0 + c_1 + c_2 + c_5 + c_6 + c_9 \tag{3.13}$$

$$s_1 = c_0 + c_2 + c_4 + c_5 + c_7 + c_8 \tag{3.14}$$

$$s_2 = c_2 + c_3 + c_4 + c_6 + c_8 + c_9 \tag{3.15}$$

$$s_3 = c_1 + c_3 + c_4 + c_5 + c_7 + c_9 \tag{3.16}$$

$$s_4 = c_0 + c_1 + c_3 + c_6 + c_7 + c_8 \tag{3.17}$$

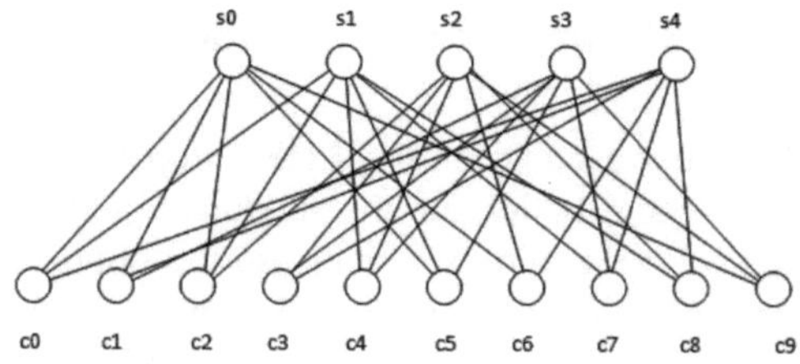

Fig. 3.1 Tanner graph

Notice any given code digit c_n is involved in exactly three syndrome digit calculations. For example, c_5 is used in Eqs. 3.13, 3.14, and 3.16 only. This stems from the fact that any given column of **H** has only three 1s. For a graphical insight to **H** and Eqs. 3.13–3.17, we may view the matrix **H** of size $(M \times N)$ as a *Tanner graph* with N *bit nodes* (also referred to as *variable nodes*) and M *check nodes* as illustrated in Fig. 3.1 for the parity-check matrix **H** in Eq. 3.11.

Alternatively, one could simply rewrite the parity-check matrix of Eq. 3.11 as follows:

$$\mathbf{H} = \begin{bmatrix} & c_0 & c_1 & c_2 & c_3 & c_4 & c_5 & c_6 & c_7 & c_8 & c_9 \\ s_0 & 1 & 1 & 1 & 0 & 0 & 1 & 1 & 0 & 0 & 1 \\ s_1 & 1 & 0 & 1 & 0 & 1 & 1 & 0 & 1 & 1 & 0 \\ s_2 & 0 & 0 & 1 & 1 & 1 & 0 & 1 & 0 & 1 & 1 \\ s_3 & 0 & 1 & 0 & 1 & 1 & 1 & 0 & 1 & 0 & 1 \\ s_4 & 1 & 1 & 0 & 1 & 0 & 0 & 1 & 1 & 1 & 0 \end{bmatrix} \tag{3.18}$$

Comparing Eq. 3.18 with Fig. 3.1, notice that the check node s_0 is connected to the six bit nodes $c_0, c_1, c_2, c_5, c_6, c_9$ corresponding to the entries in the first row of Eq. 3.18 with a number "1". Examining c_0 column of Eq. 3.18, the bit node c_0 is connected to the check nodes s_0, s_1 and s_4. Specifically, the weight of each row w_{row} (i.e., the number of 1s) is equal to 6, and the weight of each column w_{col} is equal to 3. For a practical code, $w_{row} \ll M$ and $w_{col} \ll N$ from which stems the terminology "low density." If w_{row} and w_{col} are the same for all rows and columns of **H**, the LDPC code is referred to as a *regular* LDPC code; otherwise, it's referred to as a *irregular* LDPC code. For a regular LDPC code, $\frac{w_{col}}{w_{row}} = \frac{M}{N}$ and the design code rate $R_{code} = 1 - \frac{w_{col}}{w_{row}} = 1 - \frac{M}{N}$. If after l iterations, the LDPC decoder outputs a valid codeword $\widehat{\mathbf{c}}^{(l)}$, then the syndrome

$$\mathbf{s} = [s_0, s_1, \cdots, s_{M-1}] = \widehat{\mathbf{c}}^{(l)}\mathbf{H}^T = \mathbf{0} \tag{3.19}$$

where these M syndrome bits within **s** correspond to the M check nodes. Note that a valid codeword $\widehat{\mathbf{c}}^{(l)}$ is not necessarily the transmitted codeword **c**.

3.3 Iterative Log-Likelihood LDPC Decoding Algorithm

Before we consider the LDPC decoding algorithm in detail, it is necessary to review the belief for the modulo-2 addition of two binary digits. Without any loss of generality, we shall use the opposite BPSK mapping scheme used in the previous chapter on polar codes for convenience. Thus, suppose the code digits are transmitted over an AWGN channel using BPSK in which a binary digit 0 is mapped to the signal point $-\sqrt{E_s}$ and a binary digit 1 is mapped to $\sqrt{E_s}$. Then the **reversed** log-likelihood ratio (LLR)

$$\lambda(x) = \ln\left(\frac{P(x=1)}{P(x=0)}\right) \tag{3.20}$$

- If $\lambda(x)$ is large and positive, it indicates a high confidence (belief) that $x = 1$.
- If $\lambda(x)$ is large and negative, it indicates a high confidence (belief) that $x = 0$.
- If $\lambda(x)$ is close to 0, it indicates low confidence, meaning that x is equally likely to be 0 or 1.

Note that we shall use this definition of the LLR throughout this chapter. For the modulo-2 addition of N binary random variables, let

$$y = x_1 \oplus x_2 \oplus x_3 \oplus \cdots \oplus x_N \tag{3.21}$$

$$\lambda(x_i) = \log_e\left(\frac{P(x_i=1)}{P(x_i=0)}\right) \tag{3.22}$$

$$\lambda_y = \log_e\left(\frac{P(y=1)}{1-P(y=1)}\right) \tag{3.23}$$

where $x_i \in \{0, 1\}$. Gallager [8] proved that for a sequence of N independent binary digits x_i, the probability that the whole sequence contains an even number of binary digits 1s, which is equivalent to $P(y=0)$, is given by

$$P(y=0) = \left[\frac{1}{2} + \frac{1}{2}\prod_{i=1}^{N}(1-2P(x_i=1))\right] \tag{3.24}$$

Thus, for an **odd** number of binary digits 1s

$$P(y=1) = \frac{1}{2}\left[1 - \prod_{i=1}^{N}(1-2P(x_i=1))\right] \tag{3.25}$$

Using Eq. 3.22, it can be shown that (Problem 6)

$$P\,(y=1)=\frac{1}{2}\left[1-\prod_{i=1}^{N}\tanh\left(-\frac{\lambda(x_i)}{2}\right)\right] \tag{3.26}$$

and

$$\lambda_y=\log_e\left(\frac{1-z}{1+z}\right) \tag{3.27}$$

where

$$z=\prod_{i=1}^{N}\tanh\left(-\frac{\lambda(x_i)}{2}\right) \tag{3.28}$$

Now using the relationship $\tanh\left(-\frac{u}{2}\right)=\frac{1-e^u}{1+e^u}$

$$\tanh\left(-\frac{\lambda_y}{2}\right)=\frac{1-e^{\lambda_y}}{1+e^{\lambda_y}}=\frac{1-\left(\frac{1-z}{1+z}\right)}{1+\left(\frac{1-z}{1+z}\right)}=z \tag{3.29}$$

to yield the tanh rule

$$\lambda_y=-2\tanh^{-1}\left[\prod_{i=1}^{N}\tanh\left(-\frac{\lambda(x_i)}{2}\right)\right] \tag{3.30}$$

3.3.1 Simulation: Tanh Rule for N

In simulation TanhN, a total of N binary digits are generated by a random binary source with the same probability $P\,(x_i=1)$. These digits are added modulo-2 to determine y. If $y=1$, this event is counted to estimate the probability $P\,(y=1)$ over many trials to then calculate λ_y using Eq. 3.23. This result is compared with Eq. 3.30 for $P\,(x_i=1)$ ranging from 0.1 to 0.9 for any value of N. The simulation results overlay the expected theoretical curve, as shown in Fig. 3.2 for $N=5$. For $P\,(x_i=1)=0.5$, $P\,(y=1)=0.5$ so that $\lambda_y=0$. However, for a large percentage of zeros, the modulo-2 addition of x_i is likely to be 0, which leads to a negative value of λ_y. The opposite is true as $P\,(x_i=1)$ tends toward 1.

Experiment For simplicity, each of the N binary digits was generated by a source with the same probability in this simulation. Modify the code so that each binary digit is generated with a different probability.

Fig. 3.2 Tanh rule for $N = 5$

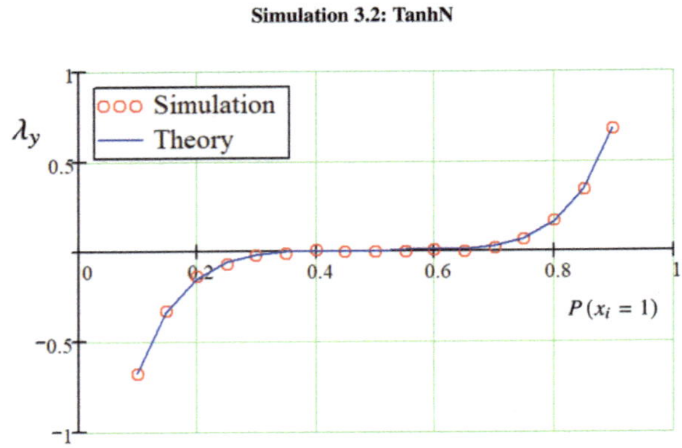

3.3.2 Extrinsic Information

Having introduced the log-likelihood ratio and the Tanh rule, consider the log-likelihood ratio $\lambda(c_n|\mathbf{r})$ given by

$$\lambda(c_n|\mathbf{r}) = \log_e \frac{P(c_n = 1|\mathbf{r})}{P(c_n = 0|\mathbf{r})} \tag{3.31}$$

Once again the sign of the log-likelihood ratio $\lambda(c_n|\mathbf{r})$ indicates the most likely value for c_n, and its magnitude is the measure of certainty. Thus, if $\lambda(c_n|\mathbf{r}) > 0$ or $\lambda(c_n|\mathbf{r}) < 0$, then it's likely that $c_n = 1$ or $c_n = 0$, respectively. If $\lambda(c_n|\mathbf{r}) \gg 0$, then its *very* likely that $c_n = 1$. Applying Bayes' rule, it can be shown for an AWGN channel (Problem 8) that

$$\lambda(c_n|\mathbf{r}) = L_c r_n + \log_e \frac{P(c_n = 1|\{r_i, i \neq n\})}{P(c_n = 0|\{r_i, i \neq n\})} \tag{3.32}$$

where L_c is referred to as the *channel reliability* defined by

$$L_c = 2\frac{\sqrt{E_b R_{code}}}{\sigma^2} \tag{3.33}$$

The quantity $(L_c r_n)$ is the *intrinsic* information, which cannot be altered for a given received noisy codeword $\mathbf{r}$. However, the *extrinsic* information $\log_e \frac{P(c_n=1|\{r_i,i\neq n\})}{P(c_n=0|\{r_i,i\neq n\})}$ can be iteratively modified to improve the log-likelihood ratio $\lambda(c_n|\mathbf{r})$ as we shall discover next.

Let $\mathbb{N}_m$ for $m = 0, 1, \ldots, M-1$ represent the set of nonzero binary digits on the mth row of $\mathbf{H}$. For example, for the matrix in Eq. 3.18, we have $\mathbb{N}_0 = \{0, 1, 2, 5, 6, 9\}, \mathbb{N}_1 = \{0, 2, 4, 5, 7, 8\}, \mathbb{N}_2 = \{2, 3, 4, 6, 8, 9\}, \mathbb{N}_3 = \{1, 3, 4, 5, 7, 9\}$, and $\mathbb{N}_4 = \{0, 1, 3, 6, 7, 8\}$. Furthermore, let $\mathbb{N}_{m,n}$ for $n = 0, 1, \ldots, N-1$ represent the set of nonzero binary digits on the mth row of $\mathbf{H}$

excluding nth column. For example, $\mathbb{N}_{0,1} = \{0, 2, 5, 6, 9\}$, $\mathbb{N}_{3,1} = \{3, 4, 5, 7, 9\}$ and $\mathbb{N}_{4,1} = \{0, 3, 6, 7, 8\}$. Finally, let $\mathbb{Z}_n$ for $n = 0, 1, \ldots, N-1$ represent the set of nonzero binary digits on the nth column of $\mathbf{H}$. For example, $\mathbb{Z}_1 = \{0, 3, 4\}$.

Let $s_{m,n}$ represent the syndrome digit calculated using the code digits connected to the check node s_m except for c_n. For example, $s_4 = c_0 + c_1 + c_3 + c_6 + c_7 + c_8$ and $s_{4,1} = c_0 + c_3 + c_6 + c_7 + c_8$ in which c_1 is excluded. If $c_1 = 1$, then $s_{4,1}$ would have to be 1 to ensure that via modulo-2 addition, $s_4 = s_{4,1} + c_1 = 0$. This in turn implies that $s_{m,1} = 1$ for all the checks in which c_1 participates so that $s_{0,1} = c_0 + c_2 + c_5 + c_6 + c_9 = 1$, $s_{3,1} = c_3 + c_4 + c_5 + c_7 + c_9 = 1$, $s_{4,1} = c_0 + c_3 + c_6 + c_7 + c_8 = 1$. Similarly, if $c_1 = 0$, then $s_{m,1} = 0$ for all the checks in which c_1 participates to ensure $s_m = 0$. Thus, for $n = 1$, given that $s_{m,1}$ should be equal to 1 for all the checks in which c_1 participates and assuming $P\left(s_{m,1} = 1 | \{r_i, i \neq 1\}\right)$ are independent, we have

$$\log_e \frac{P\left(c_1 = 1 | \{r_i, i \neq 1\}\right)}{P\left(c_1 = 0 | \{r_i, i \neq 1\}\right)} \equiv \log_e \frac{P\left(c_1 = 1 | r_0, r_2, r_3, r_4, r_5, r_6, r_7, r_8, r_9\right)}{P\left(c_1 = 0 | r_0, r_2, r_3, r_4, r_5, r_6, r_7, r_8, r_9\right)} = \tag{3.34}$$

$$\log_e \frac{P\left(s_{0,1} = 1 | \{r_i, i \neq 1\}\right) P\left(s_{3,1} = 1 | \{r_i, i \neq 1\}\right) P\left(s_{4,1} = 1 | \{r_i, i \neq 1\}\right)}{P\left(s_{0,1} = 0 | \{r_i, i \neq 1\}\right) P\left(s_{3,1} = 0 | \{r_i, i \neq 1\}\right) P\left(s_{4,1} = 0 | \{r_i, i \neq 1\}\right)} \tag{3.35}$$

$$\begin{aligned} &= \log_e \frac{P\left(s_{0,1} = 1 | \{r_i, i \neq 1\}\right)}{P\left(s_{0,1} = 0 | \{r_i, i \neq 1\}\right)} \\ &\quad + \log_e \frac{P\left(s_{3,1} = 1 | \{r_i, i \neq 1\}\right)}{P\left(s_{3,1} = 0 | \{r_i, i \neq 1\}\right)} + \log_e \frac{P\left(s_{4,1} = 1 | \{r_i, i \neq 1\}\right)}{P\left(s_{4,1} = 0 | \{r_i, i \neq 1\}\right)} \end{aligned} \tag{3.36}$$

Let's take a closer look at the first part of the above equation and expand it as shown below.

$$\log_e \frac{P\left(s_{0,1} = 1 | \{r_i, i \neq 1\}\right)}{P\left(s_{0,1} = 0 | \{r_i, i \neq 1\}\right)} = \log_e \frac{P\left(c_0 + c_2 + c_5 + c_6 + c_9 = 1 | \{r_i, i \neq 1\}\right)}{P\left(c_0 + c_2 + c_5 + c_6 + c_9 = 0 | \{r_i, i \neq 1\}\right)} \tag{3.37}$$

Now recall that the modulo-2 addition of two binary digits $\lambda(x_3) \equiv \lambda(x_1 + x_2) = \log_e \frac{P(x_1+x_2=1)}{P(x_1+x_2=0)}$. Thus, we can rewrite

$$\begin{aligned} &\log_e \frac{P\left(c_0 + c_2 + c_5 + c_6 + c_9 = 1 | \{r_i, i \neq 1\}\right)}{P\left(c_0 + c_2 + c_5 + c_6 + c_9 = 0 | \{r_i, i \neq 1\}\right)} \\ &= \lambda\left(c_0 + c_2 + c_5 + c_6 + c_9 | \{r_i, i \neq 1\}\right) \end{aligned} \tag{3.38}$$

so that Eq. 3.36 is simply

$$\log_e \frac{P\left(c_1 = 1 | \left\{r_i, i \neq 1\right\}\right)}{P\left(c_1 = 0 | \left\{r_i, i \neq 1\right\}\right)} = \lambda\left(c_0 + c_2 + c_5 + c_6 + c_9 | \left\{r_i, i \neq 1\right\}\right)$$
$$+ \lambda\left(c_3 + c_4 + c_5 + c_7 + c_9 | \left\{r_i, i \neq 1\right\}\right)$$
$$+ \lambda\left(c_0 + c_3 + c_6 + c_7 + c_8 | \left\{r_i, i \neq 1\right\}\right) \tag{3.39}$$

Using the tanh rule $\widetilde{\lambda} = -2\tanh^{-1}\left[\prod_{i=1}^{N} \tanh\left(-\frac{\lambda(x_i)}{2}\right)\right]$, we may write

$$\log_e \frac{P\left(c_1 = 1 | \left\{r_i, i \neq 1\right\}\right)}{P\left(c_1 = 0 | \left\{r_i, i \neq 1\right\}\right)}$$
$$= -2\tanh^{-1}\left(\tanh\left(-\frac{\lambda_0}{2}\right)\tanh\left(-\frac{\lambda_2}{2}\right)\tanh\left(-\frac{\lambda_5}{2}\right)\tanh\left(-\frac{\lambda_6}{2}\right)\tanh\left(-\frac{\lambda_9}{2}\right)\right)$$
$$-2\tanh^{-1}\left(\tanh\left(-\frac{\lambda_3}{2}\right)\tanh\left(-\frac{\lambda_4}{2}\right)\tanh\left(-\frac{\lambda_5}{2}\right)\tanh\left(-\frac{\lambda_7}{2}\right)\tanh\left(-\frac{\lambda_9}{2}\right)\right)$$
$$-2\tanh^{-1}\left(\tanh\left(-\frac{\lambda_0}{2}\right)\tanh\left(-\frac{\lambda_3}{2}\right)\tanh\left(-\frac{\lambda_6}{2}\right)\tanh\left(-\frac{\lambda_7}{2}\right)\tanh\left(-\frac{\lambda_8}{2}\right)\right) \tag{3.40}$$

However, $\mathbb{N}_{0,1} = \{0, 2, 5, 6, 9\}$, $\mathbb{N}_{3,1} = \{3, 4, 5, 7, 9\}$ and $\mathbb{N}_{4,1} = \{0, 3, 6, 7, 8\}$, so that

$$\log_e \frac{P\left(c_1 = 1 | \left\{r_i, i \neq 1\right\}\right)}{P\left(c_1 = 0 | \left\{r_i, i \neq 1\right\}\right)} = -2\tanh^{-1}\left(\prod_{j \in \mathbb{N}_{0,1}} \tanh\left(-\frac{\lambda_j}{2}\right)\right)$$
$$-2\tanh^{-1}\left(\prod_{j \in \mathbb{N}_{3,1}} \tanh\left(-\frac{\lambda_j}{2}\right)\right) \tag{3.41}$$
$$-2\tanh^{-1}\left(\prod_{j \in \mathbb{N}_{4,1}} \tanh\left(-\frac{\lambda_j}{2}\right)\right)$$

Now let

$$\eta_{m,n} = -2\tanh^{-1}\left(\prod_{j \in \mathbb{N}_{m,n}} \tanh\left(-\frac{\lambda_j}{2}\right)\right) \tag{3.42}$$

so that the extrinsic information for $n = 1$ is given by

$$\log_e \frac{P(c_1 = 1|\{r_i, i \neq 1\})}{P(c_1 = 0|\{r_i, i \neq 1\})} = \eta_{0,1} + \eta_{3,1} + \eta_{4,1} \tag{3.43}$$

For example, if $\eta_{4,1} > 0$, then this implies $s_{4,1} = c_0 + c_3 + c_6 + c_7 + c_8 = 1$ so that $c_1 = 1$ to ensure that $s_4 = s_{4,1} + c_1 = 0$. Recall $\mathbb{Z}_1 = \{0, 3, 4\}$, so we may further simplify this expression to

$$\log_e \frac{P(c_1 = 1|\{r_i, i \neq 1\})}{P(c_1 = 0|\{r_i, i \neq 1\})} = \sum_{m \in \mathbb{Z}_1} \eta_{m,1} \tag{3.44}$$

Therefore, for any value of n, we have

$$\lambda(c_n|\mathbf{r}) = L_c r_n + \log_e \frac{P(c_n = 1|\{r_i, i \neq n\})}{P(c_n = 0|\{r_i, i \neq n\})} = L_c r_n + \sum_{m \in \mathbb{Z}_n} \eta_{m,n} \tag{3.45}$$

where $\sum_{m \in \mathbb{Z}_n} \eta_{m,n}$ is the *total* extrinsic information on c_n.

3.3.3 Decoding Algorithm

Using the notation $\lambda_n^{[l]}$ to denote$\lambda(c_n|\mathbf{r})$ after the lth iteration,

$$\lambda_n^{[l]} = L_c r_n + \sum_{m \in \mathbb{Z}_n} \eta_{m,n}^{[l]} \tag{3.46}$$

where $\eta_{m,n}^{[l]}$ is the extrinsic information on c_n according to the mth check node on the lth iteration of the decoder given by

$$\eta_{m,n}^{[l]} = -2\tanh^{-1}\left(\prod_{j \in \mathbb{N}_{m,n}} \tanh\left(\frac{-\left(\lambda_j^{[l-1]} - \eta_{m,j}^{[l-1]}\right)}{2}\right)\right) \tag{3.47}$$

Notice that the previous value $\eta_{m,j}^{[l-1]}$ is subtracted from $\lambda_j^{[l-1]}$ to ensure that the extrinsic information is not biased. Before the first iteration of the decoder ($l = 0$), there is no extrinsic information available $\eta_{m,n}^{[0]} = 0$ so that $\lambda_n^{[0]} = L_c r_n$. After the first iteration, $\lambda_n^{[1]} = L_c r_n + \sum_{m \in \mathbb{Z}_n} \eta_{m,n}^{[1]}$, where $\eta_{m,n}^{[1]} = -2\tanh^{-1}\left(\prod_{j \in \mathbb{N}_{m,n}} \tanh\left(-\frac{\lambda_j^{[0]}}{2}\right)\right)$ because $\eta_{m,j}^{[0]} = 0$. After the second iteration,

$$\lambda_n^{[2]} = L_c r_n + \sum_{m \in \mathbb{Z}_n} \eta_{m,n}^{[2]}, \text{ where } \eta_{m,n}^{[2]} = -2 \tanh^{-1}\left(\prod_{j \in \mathbb{N}_{m,n}} \tanh\left(\frac{-\left(\lambda_j^{[1]} - \eta_{m,j}^{[1]}\right)}{2} \right) \right)$$

and so forth. After the lth iteration, if $\lambda_n^{[l]} > 0$, then we decode $\widehat{c}_n^{[l]} = 1$; otherwise, $\widehat{c}_n^{[l]} = 0$. Note that the decoder is stopped from iterating any further if a valid codeword is found, i.e., $\mathbf{s}^{[l]} = \widehat{\mathbf{c}}^{[l]}\mathbf{H}^T = \mathbf{0}$ or if $l > L_{\max}$, where $L_{\max}$ is the preset maximum number of decoder iterations.

3.4 Simple LDPC Regular Code

Using the Gauss–Jordan elimination method [10], it can be shown (Problem 4) that the generator matrix corresponding to the parity-check matrix in Eq. 3.11 is given by

$$\mathbf{G} = \begin{bmatrix} 0&1&0&1&1&1&0&0&0&0 \\ 0&0&1&1&1&0&1&0&0&0 \\ 1&0&1&0&1&0&0&1&0&0 \\ 1&1&0&1&0&0&0&0&1&0 \\ 1&1&1&0&0&0&0&0&0&1 \end{bmatrix} \tag{3.48}$$

so that $\mathbf{G}\mathbf{H}^T = \mathbf{0}$. An alternative method adopted in practice is to use the parity-check matrix $\mathbf{H}$ directly to generate the codeword. We shall consider this method later in Sect. 3.5.2 after a detailed analysis of the decoding algorithm via the following example. Suppose the message vector $\mathbf{m}$ is given by

$$\mathbf{m} = \begin{bmatrix} 1&0&0&1&1 \end{bmatrix} \tag{3.49}$$

which corresponds to the codeword $\mathbf{c}$

$$\mathbf{c} = \mathbf{m}\mathbf{G} = \begin{bmatrix} 0&1&1&0&1&1&0&0&1&1 \end{bmatrix} \tag{3.50}$$

Notice that $\mathbf{m}$ is simply the last five code digits of $\mathbf{c}$ because $\mathbf{G} = \begin{bmatrix} \mathbf{P}^T & \mathbf{I} \end{bmatrix}$. For simulation convenience, we shall convert the codeword into a column vector, i.e., $\mathbf{c} = [\mathbf{m}\mathbf{G}]^T$, in which case the syndrome $\mathbf{s}$ is given by $\mathbf{s} = \mathbf{H}\mathbf{r}$. If $\mathbf{r}$ is a valid codeword, then $\mathbf{s} = \begin{bmatrix} 0 \\ 0 \\ 0 \\ 0 \\ 0 \end{bmatrix}$.

For convenience, let's take $E_b = 8$ joules and $\sigma = \sqrt{2}$. With $R_{code} = 0.5$, we have $\frac{\sqrt{E_b R_{code}}}{\sigma^2} = 1$ so that $L_c = 2\frac{\sqrt{E_b R_{code}}}{\sigma^2} = 2$ and assume that the received noisy word is

columns 0 to 4

$$\mathbf{r}^T = \begin{bmatrix} -7.957 & -0.605 & 2.464 & 0.728 & 3.457 \end{bmatrix} \tag{3.51}$$

columns 5 to 9

$$\mathbf{r}^T = \begin{bmatrix} 3.842 & -1.666 & 0.167 & 3.966 & -1.325 \end{bmatrix} \tag{3.52}$$

Given r_n would be ideally either ± 2, its evident from simply examining $\mathbf{r}$ that r_1, r_3 and r_7 are probably errors because they are close to the value 0. Indeed, without the use of an LDPC decoder, the estimated codeword from $\mathbf{r}$ is simply $\widehat{\mathbf{c}}^T = \begin{bmatrix} 0 & \mathbf{0} & 1 & \mathbf{1} & 1 & 1 & 0 & \mathbf{1} & 1 & \mathbf{0} \end{bmatrix}$, which compared with $\mathbf{c}^T = \begin{bmatrix} 0 & 1 & 1 & 0 & 1 & 1 & 0 & 0 & 1 & 1 \end{bmatrix}$ shows that four binary digits within $\widehat{\mathbf{c}}^T$ (highlighted in bold) are incorrect. Equivalently, using only the intrinsic information $\lambda_n^{[0]} = L_c r_n$, we have

$$\mathbf{c} = \begin{bmatrix} 0 \\ 1 \\ 1 \\ 0 \\ 1 \\ 1 \\ 0 \\ 0 \\ 1 \\ 1 \end{bmatrix} \quad \lambda_n^{[0]} = \begin{bmatrix} -15.914 \\ \mathbf{-1.21} \\ 4.928 \\ \mathbf{1.456} \\ 6.914 \\ 7.684 \\ -3.332 \\ \mathbf{0.334} \\ 7.932 \\ \mathbf{-2.65} \end{bmatrix} \quad \widehat{\mathbf{c}}^{[0]} = \begin{bmatrix} 0 \\ \mathbf{0} \\ 1 \\ \mathbf{1} \\ 1 \\ 1 \\ 0 \\ \mathbf{1} \\ 1 \\ \mathbf{0} \end{bmatrix} \quad \mathbf{s}^{[0]} = \begin{bmatrix} 0 \\ 1 \\ 0 \\ 0 \\ 1 \end{bmatrix}$$

As expected, there are four binary digit errors within $\widehat{\mathbf{c}}^{[0]}$ because L_c is a constant. After the first iteration, we have three errors within $\widehat{\mathbf{c}}^{[1]}$ as follows:

$$\mathbf{c} = \begin{bmatrix} 0 \\ 1 \\ 1 \\ 0 \\ 1 \\ 1 \\ 0 \\ 0 \\ 1 \\ 1 \end{bmatrix} \quad \lambda_n^{[1]} = \begin{bmatrix} -16.403 \\ \mathbf{-3.371} \\ 6.633 \\ \mathbf{3.611} \\ 7.767 \\ 8.374 \\ -5.403 \\ -4.419 \\ 8.584 \\ \mathbf{-5.14} \end{bmatrix} \quad \widehat{\mathbf{c}}^{[1]} = \begin{bmatrix} 0 \\ \mathbf{0} \\ 1 \\ \mathbf{1} \\ 1 \\ 1 \\ 0 \\ 0 \\ 1 \\ \mathbf{0} \end{bmatrix} \quad \mathbf{s}^{[1]} = \begin{bmatrix} 0 \\ 0 \\ 0 \\ 1 \\ 0 \end{bmatrix}$$

and after the second iteration, there are still three errors within $\hat{\mathbf{c}}^{[2]}$.

$$\mathbf{c} = \begin{bmatrix} 0 \\ 1 \\ 1 \\ 0 \\ 1 \\ 1 \\ 0 \\ 0 \\ 1 \\ 1 \end{bmatrix} \quad \lambda_n^{[2]} = \begin{bmatrix} -19.277 \\ \mathbf{-4.508} \\ 7.057 \\ \mathbf{4.621} \\ 5.469 \\ 6.05 \\ -8.408 \\ -6.5 \\ 11.489 \\ \mathbf{-2.638} \end{bmatrix} \quad \hat{\mathbf{c}}^{[2]} = \begin{bmatrix} 0 \\ \mathbf{0} \\ 1 \\ \mathbf{1} \\ 1 \\ 1 \\ 0 \\ 0 \\ 1 \\ \mathbf{0} \end{bmatrix} \quad \mathbf{s}^{[2]} = \begin{bmatrix} 0 \\ 0 \\ 0 \\ 1 \\ 0 \end{bmatrix}$$

After the third iteration, a valid codeword is found, which is the same as the transmitted codeword.

$$\mathbf{c} = \begin{bmatrix} 0 \\ 1 \\ 1 \\ 0 \\ 1 \\ 1 \\ 0 \\ 0 \\ 1 \\ 1 \end{bmatrix} \quad \lambda_n^{[3]} = \begin{bmatrix} -17.488 \\ 0.737 \\ 6.309 \\ -0.919 \\ 2.713 \\ 3.443 \\ -5.579 \\ -0.889 \\ 9.555 \\ 1.256 \end{bmatrix} \quad \hat{\mathbf{c}}^{[3]} = \begin{bmatrix} 0 \\ 1 \\ 1 \\ 0 \\ 1 \\ 1 \\ 0 \\ 0 \\ 1 \\ 1 \end{bmatrix} \quad \mathbf{s}^{[3]} = \begin{bmatrix} 0 \\ 0 \\ 0 \\ 0 \\ 0 \end{bmatrix}$$

A detailed analysis of $\lambda_n^{[1]}$ will provide an insight to the operation of the decoder by example. Specifically, let us verify the value for $\lambda_7^{[1]}$ given by

$$\begin{aligned} \lambda_7^{[1]} &= \log_e \frac{P\left(c_7 = 1 | \{r_i, i \neq 7\}\right)}{P\left(c_7 = 0 | \{r_i, i \neq 7\}\right)} \\ &= L_c r_7 + \sum_{m \in \mathbb{Z}_7} \eta_{m,7}^{[1]} = L_c r_7 + \eta_{1,7}^{[1]} + \eta_{3,7}^{[1]} + \eta_{4,7}^{[1]} \end{aligned} \tag{3.53}$$

where $\eta_{m,n}^{[1]}$ for $n = 0, 1, 2, 3$ and 4 are

$$\eta_{m,n}^{[1]} = \begin{bmatrix} -0.9203 & -2.1741 & 0.9357 & 0 & 0 \\ 0.3279 & 0 & -0.3328 & 0 & -0.3286 \\ 0 & 0 & 1.1018 & 2.1664 & 1.0852 \\ 0 & -0.1786 & 0 & 0.1551 & 0.0966 \\ 0.1036 & 0.1920 & 0 & -0.1668 & 0 \end{bmatrix} \tag{3.54}$$

and $\eta_{m,n}^{[1]}$ for $n = 5, 6, 7, 8$ and 9 are

$$\eta_{m,n}^{[1]} = \begin{bmatrix} 0.9213 & -0.9999 & 0 & 0 & -1.0870 \\ -0.3282 & 0 & -4.7046 & -0.3281 & 0 \\ 0 & -1.1819 & 0 & 1.0835 & -1.2923 \\ 0.0964 & 0 & 0.5992 & 0 & -0.110 \\ 0 & 0.1113 & -0.6472 & -0.1037 & 0 \end{bmatrix} \tag{3.55}$$

Thus, $\lambda_7^{[1]} = 2(0.167) + (-4.7046 + 0.5992 - 0.6472) = 0.334 - 4.753 = -4.419$. Notice how the total extrinsic information $\sum_{m \in \mathbb{Z}_n} \eta_{m,7}^{[1]}$ of -4.753 dominates the intrinsic information $L_c r_7$ of only 0.334 to change $\lambda_7^{[0]} = 0.334$ to $\lambda_7^{[1]} = -4.419$ over the first iteration, which in turn corresponds to flipping $\hat{c}_7^{[0]} = 1$ to $\hat{c}_7^{[1]} = 0$. A closer examination of $\sum_{m \in \mathbb{Z}_n} \eta_{m,7}^{[1]}$ reveals that $\eta_{1,7}^{[1]} = -4.7046$ is the extrinsic information that dominates all the others because $\lambda_j^{[0]}$ within $\eta_{1,7}^{[1]}$ are large values. Specifically, $\eta_{1,7}^{[1]} = -2\tanh^{-1}\left(\prod_{j \in \mathbb{N}_{1,7}} \tanh\left(\frac{-\lambda_j^{[0]}}{2}\right)\right)$ where $\mathbb{N}_{1,7} = \{0, 2, 4, 5, 8\}$. Thus

$$\eta_{1,7}^{[1]} = -2\tanh^{-1}\left(\tanh\left(-\frac{\lambda_0^{[0]}}{2}\right)\tanh\left(-\frac{\lambda_2^{[0]}}{2}\right)\tanh\left(-\frac{\lambda_4^{[0]}}{2}\right)\tanh\left(-\frac{\lambda_5^{[0]}}{2}\right)\right.$$
$$\left.\tanh\left(-\frac{\lambda_8^{[0]}}{2}\right)\right) \tag{3.56}$$

$$= -2\tanh^{-1}\left(\tanh\left(\frac{-L_c r_0}{2}\right)\tanh\left(\frac{-L_c r_2}{2}\right)\tanh\left(\frac{-L_c r_4}{2}\right)\tanh\left(\frac{-L_c r_5}{2}\right)\right.$$
$$\left.\tanh\left(\frac{-L_c r_8}{2}\right)\right) \tag{3.57}$$

$$= -4.7046 \tag{3.58}$$

Notice that $s_1^{[0]} = \hat{c}_0^{[0]} + \hat{c}_2^{[0]} + \hat{c}_4^{[0]} + \hat{c}_5^{[0]} + \hat{c}_7^{[0]} + \hat{c}_8^{[0]} = 1$ and since $s_{1,7}^{[0]} = \hat{c}_0^{[0]} + \hat{c}_2^{[0]} + \hat{c}_4^{[0]} + \hat{c}_5^{[0]} + \hat{c}_8^{[0]} = 0+1+1+1+1 = 0$, this implies that $\hat{c}_7^{[1]}$ should be 0. Given that the $\lambda_j^{[0]}$ corresponding to $\mathbb{N}_{1,7} = \{0, 2, 4, 5, 8\}$ are large values, $\eta_{1,7}^{[1]}$ is a large negative number providing overwhelming evidence that $\hat{c}_7^{[1]}$ should be 0.

Thus, comparing the $n = 7$ column within $\eta_{m,n}^{[1]}$, given by $\eta_{m,7}^{[1]} = \begin{bmatrix} 0 \\ -4.7046 \\ 0 \\ 0.5992 \\ -0.6472 \end{bmatrix}$ with $\mathbf{s} = \begin{bmatrix} s_0 \\ s_1 \\ s_2 \\ s_3 \\ s_4 \end{bmatrix}$ reveals that the check nodes s_0 and s_2 provide no extrinsic information on $\widehat{c}_7^{[1]}$ because c_7 is not involved in the calculation of s_0 and s_2, Furthermore, check nodes s_1, s_3 and s_4 provide extrinsic evidence that $\widehat{c}_7^{[1]}$ should be 0, 1 and 0 respectively, with the evidence from s_1 dominating the other two. Thus, a quick glance at $\eta_{m,n}^{[l]}$ for given value of n and iteration l reveals the extrinsic information distribution due to each check node.

3.4.1 *Simulation: Rate 1/2 LDPC* (5 x 10) *Regular Code*

Simulation 3.3: LDPCDecoderSteps: The simple LDPC code presented in this section is analyzed, showing all the steps of the encoding and decoding process in great detail.
Simulation 3.4: LDPCDecoder5-10: A random message vector **m** is repeatedly sent over an AWGN channel using BPSK to find the probability of an error in a binary digit P_e versus the channel SNR (dB) to demonstrate a performance improvement by increasing $L_{\max}$ from 2 to 10.

- **Experiment:** What is the coding gain over uncoded BPSK at high SNRs? Modify the simulation to use the all-zero word $\mathbf{m} = \begin{bmatrix} 0 & 0 & 0 & 0 & 0 \end{bmatrix}$ to determine whether there is any significant change in performance.

3.5 WiMAX LDPC Codes

The WiMAX 802.16e LDPC code parity-check matrix **H** is formed from the expansion of a base model matrix $\mathbf{H}_{base}$ of size $(m_b \times n_b)$, where $n_b = 24$, $m_b = (1 - R_{code})\,24$ and R_{code} is the code rate. For example, the rate $\frac{1}{2}$ code WiMAX 802.16e standard $\mathbf{H}_{base}$ matrix is shown in Fig. 3.3.

Let $\mathbf{H}_{base}(i, j)$ represent the entry on the ith row and jth column of $\mathbf{H}_{base}$. For example, $\mathbf{H}_{base}(3, 0) = 61$ in Fig. 3.3. Each number $\mathbf{H}_{base}(i, j)$ represents a $(q$ x $q)$ matrix that is either an all-zero matrix or a circular right-shifted identity matrix,

-1	94	73	-1	-1	-1	-1	-1	55	83	-1	-1	7	0	-1	-1	-1	-1	-1	-1	-1	-1	-1	-1
-1	27	-1	-1	-1	22	79	9	-1	-1	-1	12	-1	0	0	-1	-1	-1	-1	-1	-1	-1	-1	-1
-1	-1	-1	24	22	81	-1	33	-1	-1	-1	0	-1	-1	0	0	-1	-1	-1	-1	-1	-1	-1	-1
61	-1	47	-1	-1	-1	-1	-1	65	25	-1	-1	-1	-1	-1	0	0	-1	-1	-1	-1	-1	-1	-1
-1	-1	39	-1	-1	-1	84	-1	-1	41	72	-1	-1	-1	-1	-1	0	0	-1	-1	-1	-1	-1	-1
-1	-1	-1	-1	46	40	-1	82	-1	-1	-1	79	0	-1	-1	-1	-1	0	0	-1	-1	-1	-1	-1
-1	-1	95	53	-1	-1	-1	-1	-1	14	18	-1	-1	-1	-1	-1	-1	-1	0	0	-1	-1	-1	-1
-1	11	73	-1	-1	-1	2	-1	-1	47	-1	-1	-1	-1	-1	-1	-1	-1	-1	0	0	-1	-1	-1
12	-1	-1	-1	83	24	-1	43	-1	-1	-1	51	-1	-1	-1	-1	-1	-1	-1	-1	0	0	-1	-1
-1	-1	-1	-1	-1	94	-1	59	-1	-1	70	72	-1	-1	-1	-1	-1	-1	-1	-1	-1	0	0	-1
-1	-1	7	65	-1	-1	-1	-1	39	49	-1	-1	-1	-1	-1	-1	-1	-1	-1	-1	-1	-1	0	0
43	-1	-1	-1	-1	66	-1	41	-1	-1	-1	26	7	-1	-1	-1	-1	-1	-1	-1	-1	-1	-1	0

Fig. 3.3 WiMAX 802.16e LDPC code base model matrix $\mathbf{H}_{base}$

where q is referred to as the *expansion factor* that ranges from 24, in increments of 4, to a maximum value of $q_{\max} = 96$. Thus, the parity-check matrix $\mathbf{H}$ is of size $(qm_b \times qn_b)$ with $N = qn_b$ bit nodes and $M = qm_b$ check nodes. For $R_{code} = 1/2$, $m_b = 12$, $n_b = 24$, and $q = 24$, we have the smallest codeword of length $N = 576$ code digits.

If $\mathbf{H}_{base}(i, j) < 0$, it is expanded into an all-zero $(q \times q)$ matrix. Otherwise, $\mathbf{H}_{base}(i, j)$ represents the circular right-shift of an identity matrix by amount $\left\lfloor \dfrac{q\mathbf{H}_{base}(i, j)}{q_{\max}} \right\rfloor$, where $\lfloor x \rfloor$ denotes the flooring function. When circularly shifting an identity matrix to the right, the "1" that reaches the last column is brought back to the first column of the same row. Thus, the entries in $\mathbf{H}_{base}$ are in fact the shift sizes for $q = q_{\max}$. For shorter length rate $\frac{1}{2}$ codes, notice how the shift size is scaled down by $\left\lfloor \dfrac{q\mathbf{H}_{base}(i, j)}{q_{\max}} \right\rfloor$. For example, $\mathbf{H}_{base}(3, 0) = 61$ with $q = 24$ gives $\left\lfloor \frac{24(61)}{96} \right\rfloor = 15$. The corresponding (24×24) permuted identity matrix is shown in Fig. 3.4. Given that the bottom row of an identity matrix has a "1" in its last column, shifting it 15 times would result in this "1" being placed at column 15 of that row as shown in Fig. 3.4.

3.5.1 Simulation: The $\mathbf{H}_{base}$ *Matrix*

In simulation ParityCheckMatrix, the $\mathbf{H}_{base}$ matrix with the necessary steps to determine the permutated matrix in Fig. 3.4 is presented. The $\mathbf{H}_{base}$ is then used to determine the parity-check matrix $\mathbf{H}$ for the rate 1/2 (288×576) LDPC code. You may change the expansion factor q to any of the other 18 possible values, i.e., 28, 32, 36, ..., 96.

Fig. 3.4 Permuted identity matrix corresponding to $\mathbf{H}_{base}(3, 0) = 61$ with $q = 24$

Simulation 3.5: ParityCheckMatrix

```
0 0 0 0 0 0 0 0 0 0 0 0 0 0 0 1 0 0 0 0 0 0 0 0
0 0 0 0 0 0 0 0 0 0 0 0 0 0 0 0 1 0 0 0 0 0 0 0
0 0 0 0 0 0 0 0 0 0 0 0 0 0 0 0 0 1 0 0 0 0 0 0
0 0 0 0 0 0 0 0 0 0 0 0 0 0 0 0 0 0 1 0 0 0 0 0
0 0 0 0 0 0 0 0 0 0 0 0 0 0 0 0 0 0 0 1 0 0 0 0
0 0 0 0 0 0 0 0 0 0 0 0 0 0 0 0 0 0 0 0 1 0 0 0
0 0 0 0 0 0 0 0 0 0 0 0 0 0 0 0 0 0 0 0 0 1 0 0
0 0 0 0 0 0 0 0 0 0 0 0 0 0 0 0 0 0 0 0 0 0 1 0
0 0 0 0 0 0 0 0 0 0 0 0 0 0 0 0 0 0 0 0 0 0 0 1
1 0 0 0 0 0 0 0 0 0 0 0 0 0 0 0 0 0 0 0 0 0 0 0
0 1 0 0 0 0 0 0 0 0 0 0 0 0 0 0 0 0 0 0 0 0 0 0
0 0 1 0 0 0 0 0 0 0 0 0 0 0 0 0 0 0 0 0 0 0 0 0
0 0 0 1 0 0 0 0 0 0 0 0 0 0 0 0 0 0 0 0 0 0 0 0
0 0 0 0 1 0 0 0 0 0 0 0 0 0 0 0 0 0 0 0 0 0 0 0
0 0 0 0 0 1 0 0 0 0 0 0 0 0 0 0 0 0 0 0 0 0 0 0
0 0 0 0 0 0 1 0 0 0 0 0 0 0 0 0 0 0 0 0 0 0 0 0
0 0 0 0 0 0 0 1 0 0 0 0 0 0 0 0 0 0 0 0 0 0 0 0
0 0 0 0 0 0 0 0 1 0 0 0 0 0 0 0 0 0 0 0 0 0 0 0
0 0 0 0 0 0 0 0 0 1 0 0 0 0 0 0 0 0 0 0 0 0 0 0
0 0 0 0 0 0 0 0 0 0 1 0 0 0 0 0 0 0 0 0 0 0 0 0
0 0 0 0 0 0 0 0 0 0 0 1 0 0 0 0 0 0 0 0 0 0 0 0
0 0 0 0 0 0 0 0 0 0 0 0 1 0 0 0 0 0 0 0 0 0 0 0
0 0 0 0 0 0 0 0 0 0 0 0 0 1 0 0 0 0 0 0 0 0 0 0
0 0 0 0 0 0 0 0 0 0 0 0 0 0 1 0 0 0 0 0 0 0 0 0
```

3.5.2 *WiMAX LDPC Encoder*

The message vector $\mathbf{m}^T = [m_0, m_1, \cdots, m_{M-1}]$ of size $(M \times 1)$ can be encoded via a generator transfer function matrix **G** that is determined from the parity-check matrix **H**. However, as suggested in the IEEE 802.16e™-2005 standard to lower the encoding complexity, we shall determine the codeword **c** to be transmitted directly from **H** by dividing **H** into submatrices **A**, **B**, **C**, **D**, **E**, and **T** as follows:

$$\mathbf{H} = \begin{bmatrix} \mathbf{A}\ \mathbf{B}\ \mathbf{T} \\ \mathbf{C}\ \mathbf{D}\ \mathbf{E} \end{bmatrix} \tag{3.59}$$

where the size of each matrix is given by

Matrix	*Size*
H	$M \times N$
A	$(M - q) \times M$
B	$(M - q) \times q$
T	$(M - q) \times (M - q)$
C	$q \times M$
D	$q \times q$
E	$q \times (M - q)$

(3.60)

The codeword $\mathbf{c}$ of size $(N \times 1)$ consists of $\mathbf{c} = \begin{bmatrix} \mathbf{m} \\ \mathbf{p}_1 \\ \mathbf{p}_2 \end{bmatrix}$, where $\mathbf{p}_1$ and $\mathbf{p}_2$ are parity vectors of length q and $(M - q)$ given by

$$\mathbf{p}_1 = \left(\mathbf{E}\mathbf{T}^{-1}\mathbf{A} + \mathbf{C}\right)\mathbf{m} \tag{3.61}$$

$$\mathbf{p}_2 = \mathbf{T}^{-1}\left(\mathbf{A}\mathbf{m} + \mathbf{B}\mathbf{p}_1\right) \tag{3.62}$$

Given the systematic nature of the WiMAX 802.16e LDPC code, the message vector $\mathbf{m}$ is simply the first M binary digits of the decoder estimate $\widehat{\mathbf{c}}^{(l)}$ once the decoder stops decoding.

3.5.3 Simulation: WiMAX 802.16e Rate 1/2 (288 x 576) LDPC Code

In simulation LDPCWiMAX, the performance of this LDPC code using BPSK over an AWGN channel is simulated over a range of signal-to-noise ratios. The results of this simulation are presented in Fig. 3.5 for $L_{\max} = 5$, 10, 15 and 20. As expected, the BER improves as the maximum number of iterations allowed by the decoder is increased.

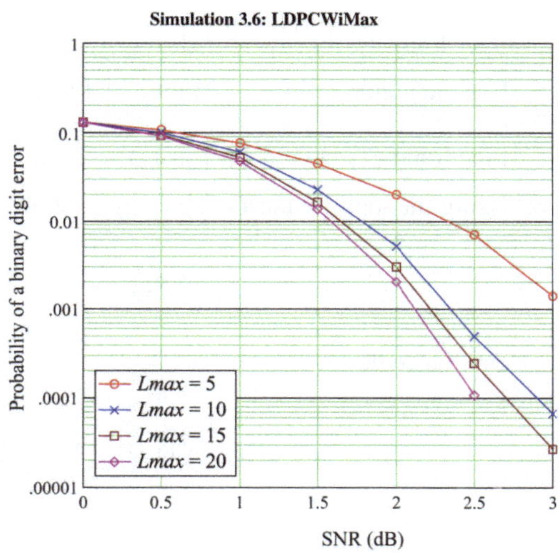

Fig. 3.5 WiMAX 802.16e rate 1/2 (288 × 576) LDPC code performance using BPSK over an AWGN channel

- **Experiment:** Increase the expansion factor q to $q_{\max} = 96$ to determine the improvement in performance by switching to the largest LDPC code within this standard.

3.6 Problems

1. Consider the following parity-check matrix $\mathbf{H} = \begin{bmatrix} 1\,0\,1\,1\,0\,0 \\ 0\,1\,1\,0\,1\,0 \\ 0\,0\,0\,1\,1\,1 \end{bmatrix}$. Determine the corresponding generator matrix $\mathbf{G}$ in systematic form $[I|P]$, where I is the identity matrix and P is a parity submatrix.
2. Determine the generator matrix $\mathbf{G}$ corresponding to the parity-check matrix $\mathbf{H} = \begin{bmatrix} 1\,1\,1\,0\,0\,0 \\ 0\,1\,1\,1\,0\,0 \\ 0\,0\,1\,1\,1\,0 \end{bmatrix}$. Check your answer using the multiplication $\mathbf{G}\mathbf{H}^\top$.
3. Verify Table 3.1, which presents the syndromes for single-digit errors in the received word.
4. Use the Gauss–Jordan elimination method to show that the generator matrix corresponding to the parity-check matrix of Eq. 3.11 is given by $\mathbf{G} = \begin{bmatrix} 0\,1\,0\,1\,1\,1\,0\,0\,0\,0 \\ 0\,0\,1\,1\,1\,0\,1\,0\,0\,0 \\ 1\,0\,1\,0\,1\,0\,0\,1\,0\,0 \\ 1\,1\,0\,1\,0\,0\,0\,0\,1\,0 \\ 1\,1\,1\,0\,0\,0\,0\,0\,0\,1 \end{bmatrix}$.
5. Use the tanh rule $\lambda_y = -2\tanh^{-1}\left[\prod_{i=1}^{N} \tanh\left(-\frac{\lambda(x_i)}{2}\right)\right]$ to derive the polar codes function $f(\lambda_1, \lambda_2) = \lambda(c_1 \oplus c_2) = 2 \cdot \tanh^{-1}\left(\tanh\left(\frac{\lambda_1}{2}\right) \cdot \tanh\left(\frac{\lambda_2}{2}\right)\right)$.
6. Using the identity $\tanh\!\left(-\frac{u}{2}\right) = \frac{1-e^u}{1+e^u}$, show from Eq. 3.22 that

$$1 - 2P(x_i = 1) = \tanh\!\left(-\frac{\lambda(x_i)}{2}\right)$$

and hence derive

$$P(y = 1) = \frac{1}{2}\left[1 - \prod_{i=1}^{N} \tanh\!\left(-\frac{\lambda(x_i)}{2}\right)\right]$$

7. Consider a scenario where three binary digits x_1, x_2, x_3 are transmitted over an AWGN channel using BPSK modulation. The received signal values are such that the corresponding LLRs are $\lambda(x_1) = 2.0$, $\lambda(x_2) = -1.5$, $\lambda(x_3) = 0.5$. Calculate the LLR λ_y for the modulo-2 addition $y = x_1 \oplus x_2 \oplus x_3$ using the tanh

rule and determine the posterior probability $P(y = 1)$ based on the computed λ_y.

8. Using Bayes theorem, show that $\lambda(c_n|\mathbf{r}) = \log_e \frac{P(c_n=1|\mathbf{r})}{P(c_n=0|\mathbf{r})} = L_c r_n + \log_e \frac{P(c_n=1|\{r_i, i\neq n\})}{P(c_n=0|\{r_i, i\neq n\})}$ for an AWGN channel where $L_c = 2\frac{\sqrt{E_b R_{code}}}{\sigma^2}$.
9. For the parity-check matrix

$$H = \begin{bmatrix} 1\,0\,1\,0\,1\,0\,1\,0 \\ 0\,1\,0\,1\,0\,1\,0\,1 \\ 0\,1\,1\,0\,1\,1\,0\,1 \\ 1\,0\,0\,1\,0\,1\,1\,0 \end{bmatrix}$$

   (a) Partition the parity-check matrix H into submatrices A, B, T, C, D and E according to the WiMAX LDPC encoding standard, using $q = 2$.
   (b) Encode the message vector

$$\mathbf{m} = [1 \quad 0 \quad 1 \quad 1]$$

   (c) Verify your codeword.
10. Using binary phase-shift keying (BPSK) to transmit LDPC code digits over an additive white Gaussian noise (AWGN) channel, derive and sketch the following expression for the channel posterior probability $P(c_n = 1|r_n) = \frac{1}{1+\exp\left[\frac{-2r_n\sqrt{E_s}}{\sigma^2}\right]}$ given that the probability density function $f(r_n|c_n = 1) = \frac{1}{\sigma\sqrt{2\pi}}\exp\left[\frac{-(r_n-\sqrt{E_s})^2}{2\sigma^2}\right]$ where c_n is the transmitted code digit and r_n is the received value at the output of the BPSK demodulator prior to any quantization.
11. The table below shows the XOR logic of binary digits.

x_1	x_2	x_3
0	0	0
0	1	1
1	0	1
1	1	0

   If we define log-likelihood ratio $\lambda(x_i) = \log_e\left(\frac{P(x_i=1)}{P(x_i=0)}\right)$ where $i = 1$ or 2, show that $P(x_3 = 1) = \frac{e^{\lambda(x_1)}+e^{\lambda(x_2)}}{(1+e^{\lambda(x_1)})(1+e^{\lambda(x_2)})}$.
12. For a rate $\frac{1}{2}$ LDPC code, using BPSK over an AWGN channel, the received word r input to the decoder is

$$r = [-7.957 \;\; -0.605 \;\; 2.464 \;\; 0.728 \;\; 3.457 \;\; 3.842 \;\; -1.666 \;\; 0.167 \;\; 3.966 \;\; -1.325]$$

and the extrinsic information $\eta_{m,n}^{[l]}$ after the first iteration is given by

$\eta_{\text{curr}} =$

−0.9203	−2.1741	0.9357	0.0000	0.0000	0.9213	−0.9999	0.0000	0.0000	−1.0870
0.3279	0.0000	−0.3328	0.0000	−0.3286	−0.3282	0.0000	−4.7046	−0.3281	0.0000
0.0000	0.0000	1.1018	2.1664	1.0852	0.0000	−1.1819	0.0000	1.0835	−1.2923
0.0000	−0.1786	0.0000	0.1551	0.0966	0.0964	0.0000	0.5992	0.0000	−0.1110
0.1036	0.1920	0.0000	−0.1668	0.0000	0.0000	0.1113	−0.6472	−0.1037	0.0000

If we assume the channel reliability $L_c = 2$, determine $\lambda_1^{[1]}$.

13. What is the decoded codeword after the first iteration given

$$\lambda_n^{[1]} = [-16.403 \; -3.371 \; 6.633 \; 3.611 \; 7.767 \; 8.374 \; -5.403 \; -4.419 \; 8.584 \; -5.14]$$

14. The parity-check matrix for a rate $\frac{1}{2}$ LDPC code is shown below.

$$\mathbf{H} = \begin{bmatrix} 1\,1\,1\,0\,0\,1\,1\,0\,0\,1 \\ 1\,0\,1\,0\,1\,1\,0\,1\,1\,0 \\ 0\,0\,1\,1\,1\,1\,0\,1\,0\,1 \\ 0\,1\,0\,1\,1\,1\,1\,0\,1\,1 \\ 1\,1\,0\,1\,0\,0\,0\,1\,1\,0 \end{bmatrix}$$

Explain why $\eta_{1,7} = \lambda((c_0 + c_2 + c_4 + c_5 + c_8) \mid r_i, i \neq 7)$, where the codeword $c = [c_0, c_1, c_2, c_3, c_4, c_5, c_6, c_7, c_8, c_9]$.

15. Using the generator matrix in the previous question, (i) if the input to the LDPC encoder is [10011], determine the corresponding codeword. (ii) Suppose this codeword is transmitted using BPSK over an AWGN channel with a channel reliability $L_c = 2$. The extrinsic information $\eta_{m,n}^{[l]}$ after the first iteration is found to be:

$\eta_{\text{curr}} =$

0	1	2	3	4	5	6	7	8	9
−0.9203	−2.1741	0.9357	0.0000	0.0000	0.9213	−0.9999	0.0000	0.0000	−1.0870
0.3279	0.0000	−0.3328	0.0000	−0.3286	−0.3282	0.0000	−4.7046	−0.3281	0.0000
0.0000	0.0000	1.1018	2.1664	1.0852	0.0000	−1.1819	0.0000	1.0835	−1.2923
0.0000	−0.1786	0.0000	0.1551	0.0966	0.0964	0.0000	0.5992	0.0000	−0.1110
0.1036	0.1920	0.0000	−0.1668	0.0000	0.0000	0.1113	−0.6472	−0.1037	0.0000

If the code digit c_7 received value is $r_7 = 0.167$, calculate $\lambda_7^{[1]}$. Highlight in your calculation which part is the intrinsic information and which specific extrinsic information dominated to correct the code digit in this iteration of the LDPC decoder.

16. What is the syndrome corresponding to

$$\lambda_n^{[1]} = [-16.403 \; -3.371 \; 6.633 \; 3.611 \; 7.767 \; 8.374 \; -5.403 \; -4.419 \; 8.584 \; -5.14]$$

17. LDPC codes suffer from performance degradation when short cycles (such as length-4 cycles) exist in their Tanner graph. Given the following parity-check matrix $\mathbf{H} = \begin{bmatrix} 1\,1\,0\,1\,0\,0 \\ 0\,1\,1\,0\,1\,0 \\ 1\,0\,1\,0\,0\,1 \end{bmatrix}$, draw the corresponding Tanner graph and identify whether there are any 4-cycles. A cycle in a Tanner graph is a path that starts at a node and returns to the same node without traversing the same edge more than once. These cycles can degrade the performance of the LDPC decoder.

Chapter 4
Turbo Codes

In this chapter, we shall consider the first significant breakthrough in the field of error-control coding. It was the tremendous success of the iterative turbo decoding algorithm that led researchers back to Low-Density Parity-Check (LDPC) codes introduced in the 1960s. Both codes are known for their near-Shannon-limit error-correcting performance. Turbo codes, by virtue of their excellent error-control capability, have been adopted for numerous applications and standards, e.g., 3G mobile telecommunications technologies, CDMA2000, digital video broadcasting (DVB-RCS), MediaFLO, and deep space communication systems.

A *turbo* code encoder typically uses two *recursive systematic convolutional* (RSC) code encoders and two *maximum-a-posteriori* (MAP) decoders to form the overall turbo decoder. Before we can delve into the encoding and decoding processes, the fundamentals of convolutional encoding and Viterbi decoding are covered in Sects. 4.1 and 4.2, respectively. Then, RSC codes are introduced in Sect. 4.3. The MAP algorithm is introduced in Sect. 4.4 before presenting the turbo code encoder in Sect. 4.5 and its corresponding decoder in Sect. 4.6. Although the MAP and turbo decoding algorithms are complex, a thorough walkthrough of a given example is provided in Sect. 4.7. Finally, the performance of the turbo decoder via simulation is presented in Sect. 4.8. The simulation code has been written to allow you to closely follow each stage of the MAP and turbo encoding and decoding processes.

4.1 Fundamentals of Convolutional Codes

A simple rate $\frac{1}{2}$ convolutional code encoder is shown in Fig. 4.1. The rectangular box represents an element of a serial shift register and $\oplus$ denotes the modulo-2 operation. For a rate 1/2 code, the *constraint length* K is the number of shifts

J. Sodha, *Fundamentals of Polar, LDPC and Turbo Codes*,
https://doi.org/10.1007/978-3-031-90924-5_4

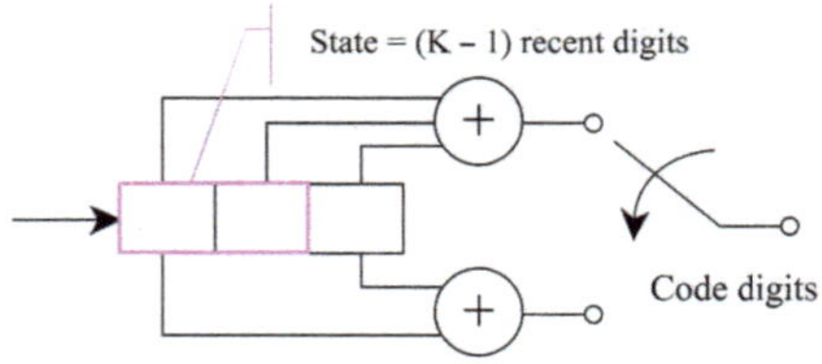

Fig. 4.1 Rate 1/2 convolutional code encoder

over which a single input digit affects the encoder output, e.g., $K = 3$ in Fig. 4.1. Furthermore, the *state* of the encoder is the content of the $K - 1$ shift registers.

In general, for a convolutional code that outputs n code digits for every k input binary digits, the code rate $R_{code} = k/n$ and the convolutional code can be described by a k rows, n columns generator transfer-function matrix $G(D)$

$$G(D) = \begin{bmatrix} G_1^{(1)}(D)\ G_1^{(2)}(D) \ldots G_1^{(n)}(D) \\ G_2^{(1)}(D) \qquad\qquad\quad G_2^{(n)}(D) \\ \vdots \qquad\qquad\qquad\qquad \vdots \\ G_k^{(1)}(D)\ G_k^{(2)}(D) \ldots G_k^{(n)}(D) \end{bmatrix} \tag{4.1}$$

where the generator polynomial in the ith row and the jth column of the matrix $G(D)$ is

$$G_i^{(j)}(D) = g_{i0}^{(j)} + g_{i1}^{(j)}D + \cdots + g_{im}^{(j)}D^m \tag{4.2}$$

where m is the *memory order* [10] of the code defined as

$$m = \max_{\substack{1\le j\le n \\ 1\le i\le k}} \left[\deg G_i^{(j)}\right] \tag{4.3}$$

For example, $G(D)$ for the rate 1/2 code shown in Fig. 4.1 is

$$G(D) = \left[G_1^{(1)}(D)\ G_1^{(2)}(D) \right] = \left[(1 + D + D^2)\ (1 + D^2) \right] \tag{4.4}$$

and the memory order $m = 2$. Refer to the problems section for more examples. The generator transfer-function matrix $G(D)$ is typically specified in either octal or vector form as illustrated in Table 4.1 in which the binary digits represent the tap connections or equivalently, $g_{im}^{(j)}$ for a given shift register. We shall use the octal representation to describe a given convolutional code. For example, the code in Fig. 4.1 would be referred to as a rate 1/2 (7, 5) code.

The *state table* for the rate 1/2 (7, 5) code is presented in Table 4.2, which lists all possible inputs to all possible initial states, with the corresponding final state and the output codeword. Using this table, the corresponding state diagram for the rate $\frac{1}{2}$, $K = 3$ encoder is shown in Fig. 4.2, in which each state is connected by a *branch*

Table 4.1 Vector or octal representation of the generator transfer-function matrix

Generator transfer-function matrix	Code Vectors	Octal form (left-justified)
$G(D) = \left[(1 + D + D^2) \ (1 + D^2) \right]$	111, 101	7,5
$G(D) = \begin{bmatrix} (1+D) & (1+D) & 1 \\ D & 0 & (1+D) \end{bmatrix}$	110, 110, 100	6, 6, 4
	010, 000, 110	2, 0, 6

Table 4.2 State table

Input	Initial state	Final state	Output codeword
0	00	00	00
1	00	10	11
0	01	00	11
1	01	10	00
0	10	01	10
1	10	11	01
0	11	01	01
1	11	11	10

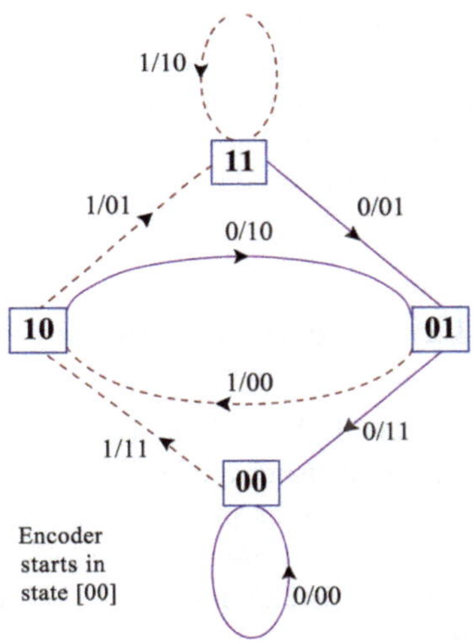

Fig. 4.2 State diagram

that is labeled with the input binary digit and the corresponding output codeword. For example, if the initial encoder state is [11] and the input binary digit is 0, then the output codeword is 01 and the new state is [01].

A code trellis is simply a state diagram redrawn. For example, the code trellis diagram corresponding to Fig. 4.2 is presented in Fig. 4.3. On this code trellis, the four possible states [00], [01], [10], [11] are labeled using the decimal equivalent representation 0, 1, 2, 3 for convenience. On the right-hand side of the code trellis, every state has two branches entering from the left-hand side. These are referred to

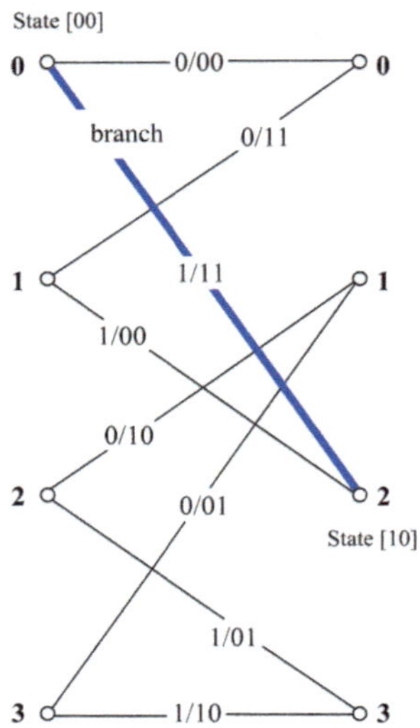

Fig. 4.3 Rate 1/2 (7, 5) code trellis

as the *upper* and *lower* branch, i.e., the state [10] has an upper branch that comes from the state [00] and a lower branch that comes from state [01]. The *branch codeword* is the codeword associated with a branch. For example, the upper branch entering state [10] has the branch codeword 11. It is labeled 1/11, which means that a binary digit 1 input to the encoder in state [00] will output the codeword 11 and move to the state [10].

4.2 Viterbi Algorithm

Andrew J. Viterbi, in [11], developed a simple maximum likelihood decoding algorithm using the serial concatenation of several code trellis diagrams referred to as a trellis, as shown in Fig. 4.4. The number of time steps in the trellis is referred to as the *trellis depth*.

For example, in Fig. 4.4, the trellis depth is 7. The time index, denoted by green numbers at the top of the diagram, runs from 0 to 7, marking each step in the decoding process. At the bottom, blue numbers represent the received sequence from the BSC, which is 11, 10, 11, 11, 01, 01, 11. These binary words are placed beneath the corresponding time steps in the code trellis. On the left-hand side of the diagram, the states are labeled 0, 1, 2, and 3 in black. The state metrics, indicated in red, begin at time index t = 0, where all states have an initial metric of zero. Let node(s, t) represent a node within this trellis, where s refers to the state (which can be 0, 1, 2, or 3) at time t.

To decide the path taken, the algorithm evaluates the branches that lead into each node. For example, at node(0, 1), there are two possible branches entering, 0/00

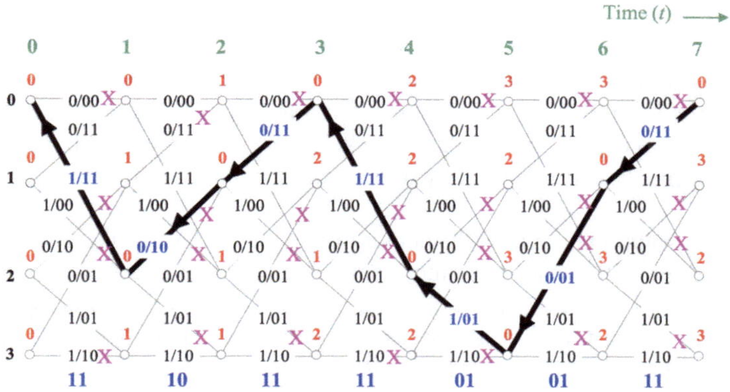

Fig. 4.4 Trellis diagram

and 0/11, where "0/00" indicates a branch with an input of 0 to the encoder and the corresponding output codeword 00. The hamming distance between the branch codeword and the received word (11, in this case) is calculated. The distances for the branches are 2 and 0, respectively. These distances are then added to the state metrics from the previous time step at node(0, 0) and node(1, 0). The branch with the lower total metric is kept, while the other is eliminated. Eliminated branches are marked by an X in the diagram. As a further example, consider node(1, 6). This node has two incoming branches, 0/10 and 0/01, with the received word being 01. The hamming distances between these branches and the received word are 2 and 0, respectively. When these distances are added to the metrics at node(2, 5) and node(3, 5), the resulting total metrics are 5 and 0. The branch with the lower total metric (in this case, 0) is kept, and the branch with the higher total is eliminated. In cases where the total metrics of two branches are equal, either branch can be selected.

The algorithm progresses through the trellis, starting from $t = 0$, with all states initially assigned zero metrics. By setting the metrics to 0 for all states, we are making no assumptions about the starting state of the encoder. However, if we know that the encoder started in the all-zero state, the metrics at $t = 0$ for states 0, 1, 2, and 3 could be set to 0, 10, 10, and 10, respectively. Any sufficiently large number, like 10, can be used to discourage paths from nonzero states. As the algorithm moves forward, the metrics at each state are updated based on the received sequence and the selected branches. By the end of the process at $t = 7$, the final state metrics for states 0, 1, 2, and 3 are 0, 3, 2, and 3, respectively. The decoding is completed by tracing back from the state with the minimum metric at $t = 7$, which in this case is state 0 with a metric of 0. This backward trace through the trellis identifies the path that corresponds to the most likely codeword sequence, which is the closest match to the received sequence.

In this specific example, no errors occurred in the received sequence, so the traceback path exactly matches the received sequence. However, if errors were present, the minimum metric at the end of the trellis would not be 0. For example, if the minimum metric was 3, this would indicate that there is a total hamming distance of 3 between the received sequence and the decoded path. Every valid path through the trellis corresponds to a possible codeword sequence that could have been generated by the encoder. By using metrics to identify the closest possible valid codeword sequence, the Viterbi algorithm achieves maximum-likelihood decoding. In this example, the decoded binary digits corresponding to the traceback path are 1001100.

For a convolutional code with memory order m, it is very likely that the 2^{km} surviving paths will merge [12] into a single path as they are traced back after a trellis of depth 5 m. Therefore, to avoid a memory-intensive and impractical large trellis depth, the starting and end points of a trellis with depth 5 m can be joined to create a circular trellis. Clever indexing is used to keep track of the two ends of the trellis as new words are inserted and decoded.

Finally, the *free distance* of a convolutional code can be used to assess its error-control capabilities [7]. It represents the minimum hamming distance between the all-zero sequence and a valid path through the trellis that begins in the all-zero state and terminates in the all-zero state. Please refer to the problems section for examples. The focus of this chapter is on Turbo codes, so we shall not cover variations such as soft-decision Viterbi decoding, punctured convolutional codes, and unequal error protection. Having presented the fundamentals of convolutional codes in brief, Sect. 4.3 marks the starting point on the path to fully understand turbo codes.

4.2.1 Simulation: Hard-Decision Viterbi Decoder

- **Simulation 4.1: ConvoEncoder**: The rate 1/2 encoder with $G(D) = \left[(1 + D + D^2)\ (1 + D^2) \right]$ is simulated.
- **Simulation 4.2: R23Encoder**: The rate 2/3 encoder with $G(D) = \begin{bmatrix} (1+D) & (1+D) & 1 \\ D & 0 & (1+D) \end{bmatrix}$ is simulated.
- **Simulation 4.3: SimpleViterbiDecoder**: A small sequence of binary digits is encoded by the rate 1/2 (7, 5) convolutional code and decoded using the Viterbi algorithm over a noiseless channel.
- In simulation VitHard, the DCS shown in Fig. 4.5 is modeled. Simulation results for the probability of an error in a binary digit P_e at the binary sink versus the channel SNR (dB) are presented in Fig. 4.6, where the coding gain at $P_e = 10^{-5}$ is approximately 1.7 dB.

  Experiment: Investigate the influence of the trellis depth on the performance of the Viterbi decoder. If you are feeling adventurous, modify the Viterbi decoder to always assume that the encoder started in the all-zero state. Does the performance improve?

Fig. 4.5 Simulated DCS: BPSK over an AWGN channel with hard-decision Viterbi decoding

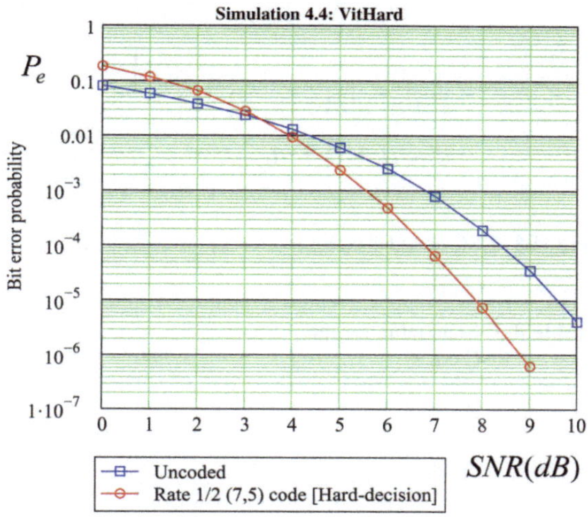

Fig. 4.6 Hard-decision Viterbi decoder performance

4.3 Recursive Systematic Convolutional Code

A rate 1/2 recursive systematic convolutional (RSC) code is shown in Fig. 4.7. The transmitted codeword is represented by $x_k = \left(x_k^{(s)}, x_k^{(p)}\right)$, where $x_k^{(s)}$ is the systematic binary digit, i.e., $x_k^{(s)} = u_k$ and $x_k^{(p)}$ is the parity binary digit. Figure 4.8 illustrates how to determine the final state and the corresponding output of the encoder. By considering all possible initial states, the state table for this RSC code is presented in Table 4.3 and its corresponding code trellis is shown in Fig. 4.9 together with the code trellis of the standard nonsystematic (NSC) rate 1/2 (7, 5) code we used in Sect. 4.1 is shown for comparison. The only difference between

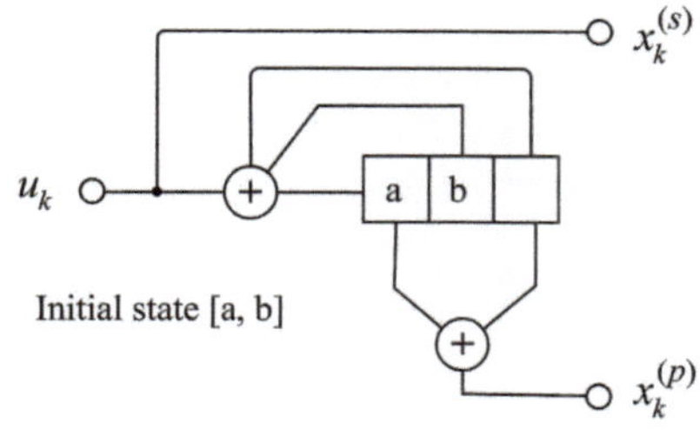

Fig. 4.7 RSC code encoder

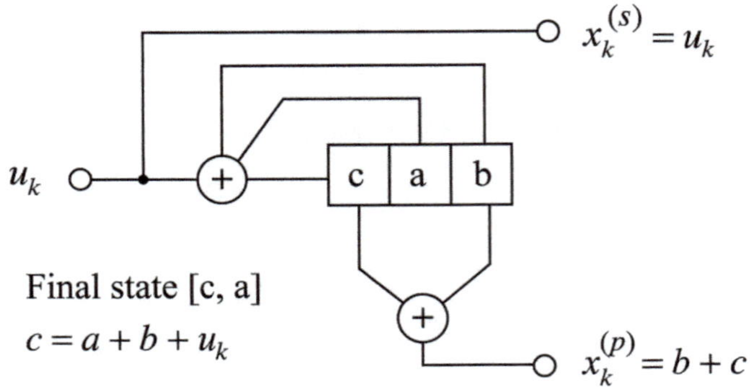

Fig. 4.8 RSC encoder operation

Table 4.3 State table for the rate 1/2 RSC code

u_k	Initial state (a, b)	Final state (c, a)	$x_k^{(s)}, x_k^{(p)}$
0	00	00	00
1	00	10	11
0	01	10	00
1	01	00	11
0	10	11	01
1	10	01	10
0	11	01	01
1	11	11	10

Fig. 4.9 Rate 1/2 recursive systematic code trellis and standard rate 1/2 trellis for comparison

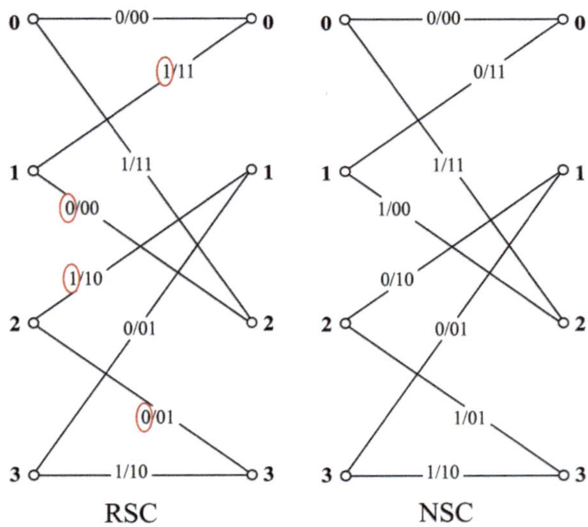

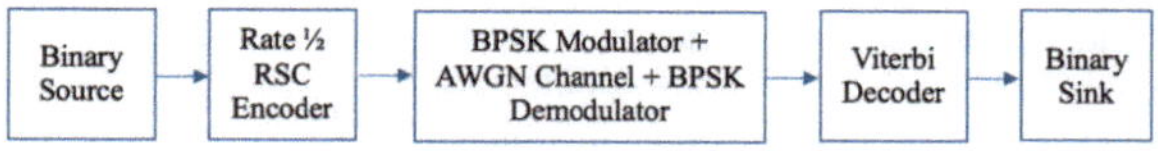

Fig. 4.10 Simulated DCS: To determine the rate 1/2 (7, 5) RSC and NSC code performance

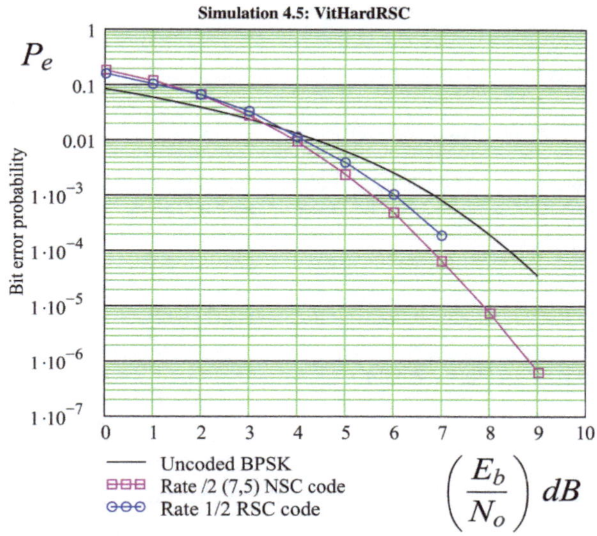

Fig. 4.11 Performance of the rate 1/2 RSC with hard-decision Viterbi decoding

these two code trellis diagrams is the information digit corresponding to a given branch codeword as highlighted.

4.3.1 Simulation: RSC Encoder with Viterbi Decoding

In simulation VitHardRSC, the DCS shown in Fig. 4.10 is modeled. The corresponding simulation results are presented in Fig. 4.11. The performance of the nonsystematic convolutional (NSC) rate 1/2 (7, 5) code is overlaid for comparison. Notice that the RSC code performs slightly better than the NSC at low SNRs. In fact, it has been shown that RSC codes with the same free distance as NSC codes generally provide a better performance at low SNRs.

4.4 MAP Algorithm

The maximum-a-posteriori (MAP) algorithm [13], which is also referred to as the BCJR algorithm—after the authors Bahl, Cocke, Jelinek, and Raviv—minimizes the bit error probability unlike the Viterbi algorithm, which is a maximum likelihood decoder. The performance of the Viterbi algorithm for most convolutional codes is

Table 4.4 MAP encoder and decoder notation

Notation	Description
u_k	Information bit input to the encoder at time k
x_k	codeword transmitted by the encoder at time k
y_k	Corresponding noisy received word at time k
$X_k^N = (x_k, x_{k+1}, \ldots, x_N)$	Transmitted code sequence from time k to time N
$Y_k^N = (y_k, y_{k+1}, \ldots, y_N)$	The noisy received sequence from time k to time N
$S_k^N = (s_k, s_{k+1}, \ldots, s_N)$	Sequence of encoder states from time k to time N, where s_k is the state of the encoder at time k, s_{k+1} is the state at time $k+1$, etc.
$\lambda_k(s) = P\left(s_k = s, Y_1^N\right)$	Probability that the encoder state is s at time k and that Y_1^N is received

virtually identical to that of the MAP decoder and its complexity is significantly lower than the MAP decoder, making it the popular choice. However, it will be shown in Sect. 4.6, that two MAP decoders, slightly modified and working together with an interleaver, form the powerful turbo decoder. We shall again use the rate 1/2 convolutional code encoder with the generator transfer matrix $G(D) = \left[(1 + D + D^2)\ (1 + D^2)\right]$ introduced in Sect. 4.1 using the notation presented in Table 4.4 to present the BCJR/MAP decoding algorithm.

For a given input to the rate 1/2 convolutional code, two additional 0s are inserted to the encoder to flush the encoder state back to the all-zero state. For example, if the input sequence u_k to the encoder is 1101, then the path through the trellis of depth 6 for this example is shown in Fig. 4.12. The state of the encoder follows the path labeled with the letters A, B, C, D, E, F, G corresponding to states 0, 2, 3, 1, 2, 1, 0. Notice how the encoder starts and ends in the all-zero state. We shall assume that this is always ensured by the encoder. Referring to Table 4.5, row A shows the information sequence with the two flush 0s highlighted in bold at time index $k = 5, 6$. The corresponding transmitted codewords are 11, 01, 01, 00, 10, 11. These codewords output by the encoder (E) are listed as decimal numbers in row B of Table 4.5. Assuming an error-free channel, the Y_1^6 received words are listed below the trellis in Fig. 4.12 and also in Table 4.5 in row C. We shall now determine the other entries in Table 4.5 that are at the heart of the MAP algorithm.

Let the probability that the encoder state is s at time k **and** the received sequence Y_1^k be denoted by

$$\alpha_k(s) = P\left(s_k = s, Y_1^k\right) \tag{4.5}$$

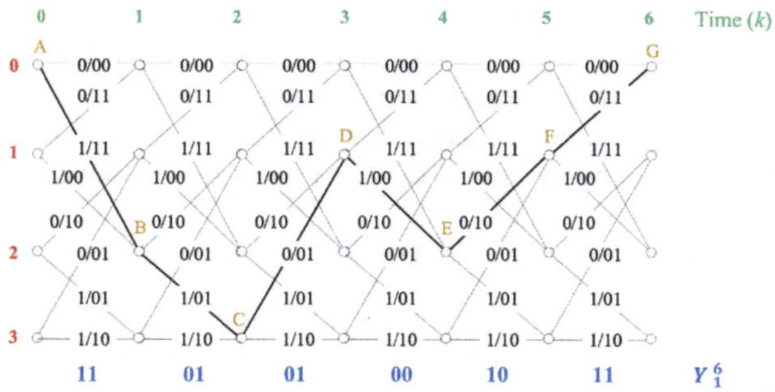

Fig. 4.12 MAP decoder trellis

Table 4.5 MAP decoding example under a noiseless channel

	k	0	1	2	3	4	5	6
A	u_k		1	1	0	1	**0**	**0**
B	E		3	1	1	0	2	3
C	Y_1^6		11	01	01	00	10	11
D	$\alpha_k(0)$	1	0	0	0	0	0	0.016
E	$\alpha_k(1)$	0	0	0	0.125	0	0.031	0
F	$\alpha_k(2)$	0	0.5	0	0	0.063	0	0
G	$\alpha_k(3)$	0	0	0.25	0	0	0	0
H	$\beta_k(0)$	0.016	0	0	0	0	0	1
I	$\beta_k(1)$	0	0	0	0.125	0	0.5	0
J	$\beta_k(2)$	0	0.031	0	0	0.25	0	0
K	$\beta_k(3)$	0	0	0.063	0	0	0	0
L	$\lambda_k(0)$	0.016	0	0	0	0	0	0.016
M	$\lambda_k(1)$	0	0	0	0.016	0	0.016	0
N	$\lambda_k(2)$	0	0.016	0	0	0.016	0	0
O	$\lambda_k(3)$	0	0	0.016	0	0	0	0
P	$P\left(u_k = 0 \mid Y_1^N\right)$		0	0	1	0	1	1
Q	$\widehat{u_k}$		1	1	0	1	**0**	**0**

where $\alpha_0(0) = 1$ because encoder started in the all-zero state. It can be shown [13] that the probabilities $\alpha_k(s)$ are given by

$$\alpha_k(s) = \sum_{s' \in S} \alpha_{k-1}(s')\gamma_k(s', s) \tag{4.6}$$

where $\gamma_k(s', s) = P\left(s_k = s, y_k | s_{k-1} = s'\right)$ is the probability that the encoder state is s at time k and y_k is the received word given the previous state at time $k-1$ is s'. Taking a closer look at this equation, notice that the sum $s' \in S$ is taken over all possible states s' at the previous time index $k-1$. This accounts for all the ways the encoder could have transitioned from any previous state s' to the current state s at time k. The $\alpha_{k-1}(s')$ represents the joint probability of being in state s' at time $k-1$ and the product $\alpha_{k-1}(s')\gamma_k(s', s)$ calculates the probability of being in state s' at time $k-1$ and then transitioning to state s at time k. By summing this product over all possible previous states, the equation effectively accumulates the total probability of arriving at state s at time k from any of the possible previous states.

For $k \geq 1$ and $s = 0$, $\alpha_k(0) = \sum_{s' \in S} \alpha_{k-1}(s')\gamma_k(s', 0)$ expands to $\alpha_{k-1}(0)\gamma_k(0, 0) + \alpha_{k-1}(1)\gamma_k(1, 0) + \alpha_{k-1}(2)\gamma_k(2, 0) + \alpha_{k-1}(3)\gamma_k(3, 0)$. However, $\gamma_k(2, 0) = \gamma_k(3, 0) = 0$ because these transitions are not possible. Repeating this process for all values of s, we find at time index k

$$\alpha_k(0) = \alpha_{k-1}(0)\gamma_k(0, 0) + \alpha_{k-1}(1)\gamma_k(1, 0) \tag{4.7}$$

$$\alpha_k(1) = \alpha_{k-1}(2)\gamma_k(2, 1) + \alpha_{k-1}(3)\gamma_k(3, 1) \tag{4.8}$$

$$\alpha_k(2) = \alpha_{k-1}(0)\gamma_k(0, 2) + \alpha_{k-1}(1)\gamma_k(1, 2) \tag{4.9}$$

$$\alpha_k(3) = \alpha_{k-1}(2)\gamma_k(2, 3) + \alpha_{k-1}(3)\gamma_k(3, 3) \tag{4.10}$$

To determine $\gamma_k(s', s)$, it can be shown (Problem 10) that

$$\gamma_k(s', s) = P\left(s_k = s | s_{k-1} = s'\right) P\left(y_k | b_k\right) \tag{4.11}$$

where b_k is the branch codeword on the code trellis diagram between states s' and s. If the input binary digit u_k is 0 or 1 with a probability 0.5, then $P\left(s_k = s | s_{k-1} = s'\right) = \frac{1}{2}$. Let h represent the hamming distance between y_k and b_k. If the codewords are transferred over a BSC with a crossover probability α_{bsc}, then $P\left(y_k | b_k\right) = \alpha_{bsc}^h \left(1 - \alpha_{bsc}\right)^{2-h}$ and therefore

$$\gamma_k(s', s) = \frac{\alpha_{bsc}^h (1 - \alpha_{bsc})^{2-h}}{2} \tag{4.12}$$

For a noiseless channel where $\alpha_{bsc} = 0$, $\gamma_k(s', s) = \frac{1}{2}$. Accordingly, for example, $\alpha_1(2) = \alpha_0(0)\gamma_k(0, 2) + \alpha_0(1)\gamma_k(1, 2) = \gamma_k(0, 2) = \frac{1}{2}$ because $\alpha_0(1) = 0$ and $\alpha_0(0) = 1$. Repeating this calculation for all values of k and s, we find the only nonzero $\alpha_k(s)$ probability at time k is given by $(0.5)^k$ as shown in Table 4.5 rows D–G. This is only half of the MAP algorithm operation. The next phase to calculate $\beta_k(s) = P\left(Y_{k+1}^N | s_k = s\right)$, which is the probability of observing the received sequence from time $k+1$ to N given the state at time k is s. Now $\beta_N(0) = 1$ because the encoder state is forced to end in the all-zero state. It can be shown [13] that

$$\beta_{k-1}(s') = \sum_{s \in S} \beta_k(s)\gamma_k(s', s) \tag{4.13}$$

Each term in the sum represents the contribution of a transition from state s' at time $k-1$ to state s at time k, weighted by the probability of such a transition $\gamma_k(s', s)$ and the likelihood of the subsequent sequence $\beta_k(s)$. The equation performs a backward recursion, starting from the known final state $s_N = 0$ so that $\beta_N(0) = 1$ and moving backward through the trellis to compute the backward variables for earlier time steps. Thus, for the trellis in Fig. 4.12

$$\beta_{k-1}(0) = \beta_k(0)\gamma_k(0, 0) + \beta_k(2)\gamma_k(0, 2) \tag{4.14}$$

$$\beta_{k-1}(1) = \beta_k(0)\gamma_k(1, 0) + \beta_k(2)\gamma_k(1, 2) \tag{4.15}$$

$$\beta_{k-1}(2) = \beta_k(1)\gamma_k(2, 1) + \beta_k(3)\gamma_k(2, 3) \tag{4.16}$$

$$\beta_{k-1}(3) = \beta_k(1)\gamma_k(3, 1) + \beta_k(3)\gamma_k(3, 3) \tag{4.17}$$

These $\beta_k(s)$ probabilities for a noiseless channel are presented in Table 4.5 rows H to K. In rows L–O is the probability $\lambda_k(s) = P(s_k = s, Y_1^N)$ that the encoder state is s at time k and Y_1^N is received, which can be shown (Problem 11) to be given by

$$\lambda_k(s) = \alpha_k(s)\beta_k(s) \tag{4.18}$$

and

$$P\left(u_k = 0|Y_1^N\right) = \frac{1}{\lambda_N(0)}\sum_{S^-}\lambda_k(s) \tag{4.19}$$

where S^- is the set of ordered pairs $\left(s, s'\right)$ corresponding to all the state transitions $\left(s_{k-1} = s'\right) \rightarrow (s_k = s)$ caused by the input bit $u_k = 0$. The sum $\sum_{S^-}\lambda_k(s)$ adds up the probabilities of all such paths through the trellis diagram that are consistent with $u_k = 0$ and then normalized with $\lambda_N(0)$ to allow the estimate $\widehat{u_k} = 0$ if $P\left(u_k = 0|Y_1^N\right) \geq 0.5$, otherwise $\widehat{u_k} = 1$. Referring to the trellis in Fig. 4.12, $\sum_{S^-}\lambda_k(s) = \lambda_k(0) + \lambda_k(1)$. For example, for $k = 3$, $\lambda_3(0) + \lambda_3(1) = 0 + 0.016 = 0.016$. Given that $\lambda_6(0) = 0.016$, we find $P\left(u_3 = 0|Y_1^N\right) = \frac{0.016}{0.016} = 1$, which is greater than 0.5, thus $\widehat{u_3} = 0$. Table 4.5 shows all the decoding decisions in row P, which as expected corresponds to row A because it's a noiseless channel.

4.4.1 Simulation: Map Decoder

Simulation 4.6: MapDecoderSteps: All the steps of the noiseless MAP decoding example are presented in this simulation.

Fig. 4.13 Simulated DCS: BPSK over an AWGN with hard-decision MAP decoding

Fig. 4.14 MAP decoder BER

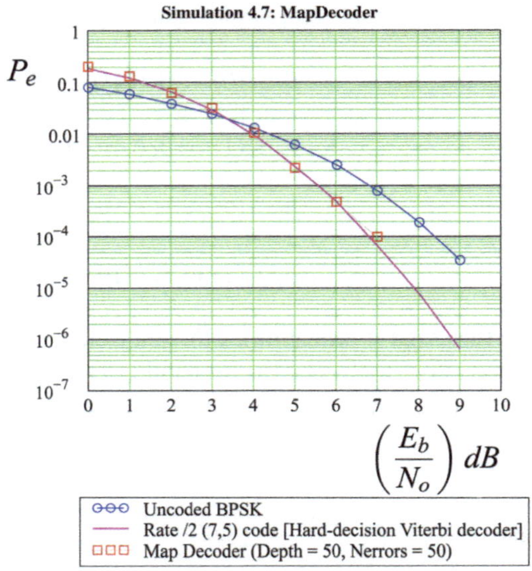

- **Experiment:** Change the BSC crossover probability α_{bsc} to a nonzero value.

In simulation MapDecoder, the DCS shown in Fig. 4.13 is modeled. The corresponding simulation results are presented in Fig. 4.14. Although there is virtually no difference between the performance of the MAP decoder and the corresponding Viterbi decoder, there is an important difference between them as follows. The Viterbi algorithm minimizes the sequence error probability by finding the most likely path through the trellis, whereas the MAP decoder focuses on the most likely value of each individual bit u_k.

- **Experiment:** Observe the performance variation as the trellis depth is increased from 6 to 50. For a small trellis depth of 6, an improved performance is evident because the impact of the two flushing 0s—which reduce information throughput—has not been taken into account within the code rate. For a trellis depth of 6, the code rate $R_{code} = 1/3$ because 12 code digits are sent for every four information digits. Update the simulation code to set $R_{code} = \frac{n}{2n+4}$ where n is the number of information binary digits within the trellis depth N.

4.5 Turbo Code Encoder

A turbo code encoder consists of the parallel concatenation of two or more, usually identical, recursive systematic convolutional (RSC) code encoders and a M_I x M_I nonuniform interleaver. Figure 4.15 illustrates a rate 1/2 turbo encoder where u_k is the information bit input to the first encoder E1 at time k, and u_k^* represents the interleaved information bit input to the second encoder E2. We shall represent the interleaving and de-interleaving process by $I[.]$ and $D[.]$, respectively, so that we may write $u_k^* = u_{I[k]}$. For a given RSC encoder (E1 or E2), let $x_k = \left(x_k^{(s)}, x_k^{(p)}\right)$ represent the transmitted codeword and let $y_k = \left(y_k^{(s)}, y_k^{(p)}\right)$ represent the noisy received word by the corresponding modified MAP decoder, where $y_k^{(s)}$ is the noisy received systematic digit and $y_k^{(p)}$ is the noisy received parity bit. Specifically, for the encoder shown in Fig. 4.15, $\left(x_k^{(1s)}, x_k^{(1p)}\right)$ and $\left(x_k^{(2s)}, x_k^{(2p)}\right)$, respectively, represent the codewords generated by the encoders E1 and E2, where $x_k^{(1s)} = u_k$ and $x_k^{(2s)} = u_{I[k]}$. By alternatively puncturing the parity code digits, the overall rate of the turbo encoder is increased from 1/3 to 1/2, i.e., the transmitted codeword sequence is $x_1^{(1s)}x_1^{(1p)}, x_2^{(1s)}x_2^{(2p)}, x_3^{(1s)}x_3^{(1p)}, x_4^{(1s)}x_4^{(2p)}, \ldots$.

Using the BPSK modulation scheme over an AWGN, the demodulator is configured such that the signal constellation coordinate is ideally $+1$ or -1 if the transmitted code digit is either a binary digit 1 or 0, respectively. In this case, the unknown punctured code digit value can be conveniently set to 0 without any bias, i.e., $y_1^{(1s)}y_1^{(1p)}0,\ y_2^{(1s)}0y_2^{(2p)},\ y_3^{(1s)}y_3^{(1p)}0,\ y_4^{(1s)}0y_4^{(2p)}, \cdots$ We shall come back to this point when we consider the decoder in the next section.

The function of the interleaver is to permute the information sequence so that the two encoders operate on the same set of information binary digits, but in different order. We shall consider the interleaver as presented in Berrou and Glavieux [14]. Let the function *mod* $[a, b]$ represent the modulus of a with b, which calculates the remainder when one number is divided by another. For example, *mod* $[4, 2] = 0$, *mod* $[4, 3] = 1$, *mod* $[4, 5] = 4$. Let i and j be the row and column position of a square matrix of size M_I x M_I for *writing*, and let i_{row} and j_{col} for *reading*. For

Fig. 4.15 Rate 1/2 turbo encoder

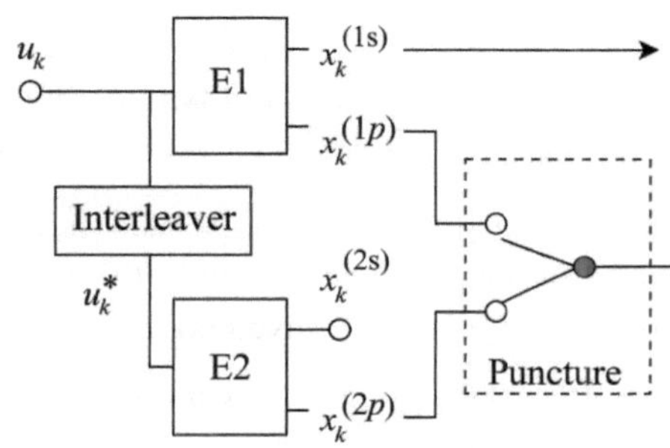

Fig. 4.16 Original square matrix for $M_I = 8$

$$\begin{pmatrix} 0 & 1 & 2 & 3 & 4 & 5 & 6 & 7 \\ 8 & 9 & 10 & 11 & 12 & 13 & 14 & 15 \\ 16 & 17 & 18 & 19 & 20 & 21 & 22 & 23 \\ 24 & 25 & 26 & 27 & 28 & 29 & 30 & 31 \\ 32 & 33 & 34 & 35 & 36 & 37 & 38 & 39 \\ 40 & 41 & 42 & 43 & 44 & 45 & 46 & 47 \\ 48 & 49 & 50 & 51 & 52 & 53 & 54 & 55 \\ 56 & 57 & 58 & 59 & 60 & 61 & 62 & 63 \end{pmatrix}$$

Fig. 4.17 Interleaved square matrix for $M_I = 8$

Simulation 4.8: Interleaver

$$\begin{pmatrix} 0 & 41 & 16 & 59 & 36 & 9 & 50 & 31 \\ 44 & 21 & 62 & 35 & 10 & 53 & 24 & 7 \\ 18 & 57 & 34 & 11 & 48 & 25 & 6 & 47 \\ 60 & 33 & 12 & 51 & 26 & 5 & 42 & 23 \\ 32 & 13 & 54 & 27 & 4 & 45 & 20 & 63 \\ 14 & 49 & 28 & 3 & 40 & 17 & 58 & 39 \\ 52 & 29 & 2 & 43 & 22 & 61 & 38 & 15 \\ 30 & 1 & 46 & 19 & 56 & 37 & 8 & 55 \end{pmatrix}$$

nonuniform interleaving

$$i_{row} = \text{mod}\left[\left(\frac{M_I}{2} + 1\right)(i + j), M_I\right] \tag{4.20}$$

$$j_{col} = \text{mod}\,[(j + 1)P_x - 1, M_I] \tag{4.21}$$

where P_x is a number relatively prime with M_I in the array $P_0 = 17$, $P_1 = 37$, $P_2 = 19$, $P_3 = 29$, $P_4 = 41$, $P_5 = 23$, $P_6 = 13$, $P_7 = 7$ with index x given by $x = mod\,[(i + j), 8]$. For example, for $M_I = 8$, Figs. 4.16 and 4.17 illustrate the result of interleaving a square matrix, which contains the numbers 0, 1, 2, ..., 63. This interleaving process, which is valid for $M_I = 8, 16, 32, 64, 128, 256$, etc., is only one of many other possible variations.

4.6 Turbo Decoder

The turbo decoder consists of two modified MAP decoders as shown in Fig. 4.18. Recall that $\left(x_k^{(1s)}, x_k^{(1p)}\right)$ and $\left(x_k^{(2s)}, x_k^{(2p)}\right)$, respectively, represent the codewords generated by the encoders E1 and E2, where $x_k^{(1s)} = u_k$ and $x_k^{(2s)} = u_k^*$. The subscript symbols "s" and "p" refer to the systematic and parity bit. The corresponding noisy received codewords are $\left(y_k^{(1s)}, y_k^{(1p)}\right)$ and $\left(y_k^{(2s)}, y_k^{(2p)}\right)$. For example, since $x_k^{(2s)} = u_k^*$ is not transmitted, $y_k^{(2s)}$ is determined by interleaving $y_k^{(1s)}$, or equivalently $y_k^{(2s)} = y_{I[k]}^{(1s)}$. As before, let $X_k^N = (x_k, x_{k+1}, \ldots, x_N)$ represent the codeword sequence generated by the RSC encoder from time k to

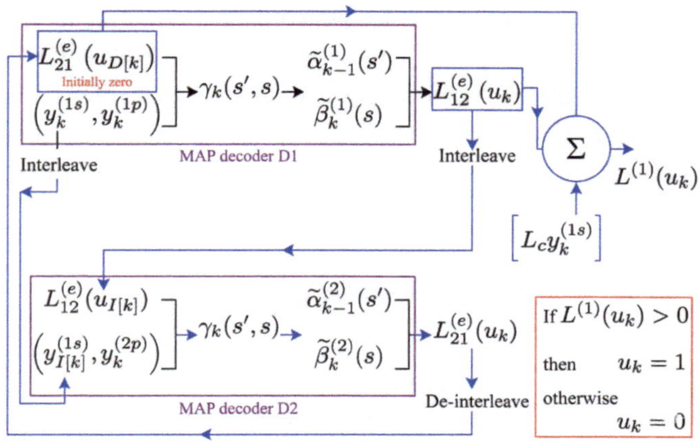

Fig. 4.18 Rate 1/2 turbo decoder operation

N, and let $Y_k^N = (y_k, y_{k+1}, \ldots, y_N)$ represent the corresponding noisy received codeword sequence. If $S_k^N = (s_k, s_{k+1}, \ldots, s_N)$ represents the sequence of encoder states from time k to N, where s_k is the state of an RSC code encoder at time k, then as in Ryan [15], we define the following modified probabilities

$$\widetilde{\alpha}_k(s) = \frac{\alpha_k(s)}{P\left(Y_1^k\right)} \text{ and } \widetilde{\beta}_k(s) = \frac{\beta_k(s)}{P\left(Y_{k+1}^N | Y_1^k\right)} \tag{4.22}$$

where (Problem 12) the summation $\sum_s \widetilde{\alpha}_k(s) = 1$. It can be shown [15] that the modified probabilities $\widetilde{\alpha}_k(s)$ and $\widetilde{\beta}_{k-1}^{(1)}(s')$ for a given MAP decoder (D1 or D2) for $k \geq 1$ are given by

$$\widetilde{\alpha}_k(s) = \frac{\sum_{s'} \widetilde{\alpha}_{k-1}(s')\gamma_k(s', s)}{\sum_s \sum_{s'} \widetilde{\alpha}_{k-1}(s')\gamma_k(s', s)} \tag{4.23}$$

and

$$\widetilde{\beta}_{k-1}(s') = \frac{\sum_{s} \widetilde{\beta}_k(s)\gamma_k(s', s)}{\sum_s \sum_{s'} \widetilde{\alpha}_{k-1}(s')\gamma_k(s', s)} \tag{4.24}$$

where $\gamma_k(s', s) = P\left(s_k = s, y_k | s_{k-1} = s'\right)$ is given by

$$\gamma_k(s', s) = \gamma_k^{(e)}(s', s) \exp\left[\frac{u_k}{2}\left(L^{(e)}(u_k) + L_c y_k^{(s)}\right)\right] \tag{4.25}$$

In this expression for $\gamma_k(s', s)$, u_k is the first code digit of the branch codeword between $\left(s_{k-1} = s'\right) \longrightarrow (s_k = s)$. This is because for an RSC encoder, u_k for D1 and D2 is either $x_k^{(1s)}$ or $x_k^{(2s)}$ respectively, and $y_k^{(s)}$ for D1 and D2 is either $y_k^{(1s)}$ or $y_k^{(2s)}$, respectively. Furthermore, $L^{(e)}(u_k)$ is $L_{21}^{(e)}(u_k)$ for the extrinsic information from D2 to D1 and $L_{12}^{(e)}(u_k)$ for the extrinsic information from D1 to D2. Lastly, L_c is given by

$$L_c = 4R_{code}\left(\frac{E_b}{N_o}\right) \tag{4.26}$$

where R_{code} is the code rate, and the ratio $\frac{E_b}{N_o}$ is given by $\frac{E_b}{N_o} = 10^{\frac{SNR}{10}}$, where SNR is the channel signal-to-noise ratio expressed in decibels. The individual terms of foregoing equations are summarized in Table 4.6 for decoders D1 and D2 for convenience.

If the code digits are transferred over the channel using BPSK with coherent demodulation, then the punctured code digits can be set to 0. The operation of the turbo decoder is presented in Fig. 4.18. Initially, the deinterleaved extrinsic information $L_{21}^{(e)}\left(u_{D[k]}\right)$ passed from D2 to D1 is 0. Using the noisy codeword

Table 4.6 Summary of Turbo decoding equations

MAP decoder D1	MAP decoder D2
$\gamma_k^{(e)}(s', s) =$ $\gamma_k^{(1e)}(s', s) = \exp\left[\frac{L_c y_k^{(1p)} x_k^{(1p)}}{2}\right]$	$\gamma_k^{(e)}(s', s) =$ $\gamma_k^{(2e)}(s', s) = \exp\left[\frac{L_c y_k^{(2p)} x_k^{(2p)}}{2}\right]$
Extrinsic information from D2 to D1 $L^{(e)}(u_k) = L_{21}^{(e)}(u_k)$	Extrinsic information from D1 to D2 $L^{(e)}(u_k) = L_{12}^{(e)}(u_k)$
$y_k^{(s)} = y_k^{(1s)}$	$y_k^{(s)} = y_k^{(2s)}$
$\gamma_k(s', s) = \gamma_k^{(e)}(s', s)\exp\left[\frac{u_k}{2}\left(L^{(e)}(u_k) + L_c y_k^{(s)}\right)\right]$	
$L_{21}^{(e)}(u_k) = \log\left(\frac{\sum_{S+} \widetilde{\alpha}_{k-1}^{(2)}(s')\gamma_k^{(2e)}(s',s)\widetilde{\beta}_k^{(2)}(s)}{\sum_{S-} \widetilde{\alpha}_{k-1}^{(2)}(s')\gamma_k^{(2e)}(s',s)\widetilde{\beta}_k^{(2)}(s)}\right)$ where $\widetilde{\alpha}_{k-1}^{(2)}(s')$ and $\widetilde{\beta}_k^{(2)}(s)$ are the modified probabilities corresponding to D2.	
$L_{12}^{(e)}(u_k) = \log\left(\frac{\sum_{S+} \widetilde{\alpha}_{k-1}^{(1)}(s')\gamma_k^{(1e)}(s',s)\widetilde{\beta}_k^{(1)}(s)}{\sum_{S-} \widetilde{\alpha}_{k-1}^{(1)}(s')\gamma_k^{(1e)}(s',s)\widetilde{\beta}_k^{(1)}(s)}\right)$ where $\widetilde{\alpha}_{k-1}^{(1)}$ and $\widetilde{\beta}_k^{(1)}(s)$ are the modified probabilities corresponding to D1. S^+ and S^- are the set of state transitions $\left(s_{k-1} = s'\right) \longrightarrow (s_k = s)$ that are the result of an input information digit '1' or '0' respectively to the encoder at time k.	

$\left(y_k^{(1s)}, y_k^{(1p)}\right)$, the $\gamma_k(s', s)$ probabilities for D1 are calculated. These are then used to determine $\widetilde{\alpha}_{k-1}^{(1)}$ and $\widetilde{\beta}_k^{(1)}(s)$, which in turn are used to determine the extrinsic information $L_{12}^{(e)}(u_k)$ to be passed from D1 to D2. This extrinsic information is first interleaved to create $L_{12}^{(e)}\left(u_{I[k]}\right)$, and $y_k^{(1s)}$ is interleaved to create $y_{I[k]}^{(1s)}$.

Using $L_{12}^{(e)}\left(u_{I[k]}\right)$ and $\left(y_{I[k]}^{(1s)}, y_k^{(2p)}\right)$, the $\gamma_k(s', s)$ probabilities for D2 are calculated. These are then used to determine $\widetilde{\alpha}_{k-1}^{(2)}(s')$ and $\widetilde{\beta}_k^{(2)}(s)$, which in turn are used to determine the extrinsic information $L_{21}^{(e)}(u_k)$ to be passed from D2 to D1. This extrinsic information is then deinterleaved to create $L_{21}^{(e)}\left(u_{D[k]}\right)$, marking the completion of first iteration of the turbo decoder. At this stage, the log-likelihood ratio $L^{(1)}(u_k)$ for D1 is computed and is given by

$$L^{(1)}(u_k) = L_c y_k^{(1s)} + L_{21}^{(e)}\left(u_{D[k]}\right) + L_{12}^{(e)}(u_k) \tag{4.27}$$

At time k, the estimate $\widehat{u_k} = 1$ if $L^{(1)}(u_k) > 0$, otherwise $\widehat{u_k} = 0$. This first iteration estimate of the information sequence $\{u_k\}$ can be improved by repeating the calculation cycle to determine the second iteration estimate and so forth as summarized in Fig. 4.18. The quantity $L_c y_k^{(1s)}$, referred to as the *channel value*, is dictated by the channel noise and remains constant for a given time k throughout the iteration cycle.

4.7 Decoding Example

To illustrate the turbo decoding algorithm step by step, let the information sequence to be encoded be 1100100100100110 as shown in row A of Table 4.7. The first step is to initialize the turbo decoder by setting within D1, the modified probabilities $\widetilde{\alpha}_0^{(1)}(0) = 1$ and $\widetilde{\alpha}_0^{(1)}(s) = 0$ for $s \neq 0$. Similarly, $\widetilde{\beta}_N^{(1)}(0) = 1$, $\widetilde{\beta}_N^{(1)}(s) = 0$ for $s \neq 0$ and $L_{21}^{e}(u_k) = 0$ for $k = 1, 2, \ldots, N$. Within D2, $\widetilde{\alpha}_0^{(2)}(0) = 1$ and $\widetilde{\alpha}_0^{(2)}(s) = 0$ for $s \neq 0$. However, $\widetilde{\beta}_N^{(2)}(s) = \widetilde{\alpha}_N^{(2)}(s)$ after $\widetilde{\alpha}_N^{(2)}(s)$ is computed in the first iteration and $L_{12}^{e}(u_k)$ will be determined from D1 after the first half-iteration. These steps will be shown in detail in this example.

Row B lists the corresponding encoder E1 output using the rate 1/2 recursive systematic convolutional (RSC) code in Fig. 4.7. Using its code trellis in Fig. 4.9a, the output codewords are 11, 10, 00, 01, 10, …, which are represented by a decimal number in row B as 3, 2, 0, 1, 2, etc. For $M_I = 4$, the interleave Eq. 4.21, for example, is **not** valid. However, let us interleave the ordered index 1, 2, 3, 4,..., 16 to 1, 14, 9, 8, 15, 10, 7, 4, 11, 6, 3, 16, 5, 2, 13, 12. Accordingly, u_k in row A is interleaved to produce $u_{I[k]}$ shown in row C. These $u_{I[k]}$ bits are input to the encoder E2 to produce row D. For a given input bit u_k to the turbo encoder, the transmitted code digits are either the pair $\left(x_k^{(1s)}, x_k^{(1p)}\right)$ or $\left(x_k^{(1s)}, x_k^{(2p)}\right)$.

Table 4.7 Turbo decoding example at SNR = 0 dB

	k	0	1	2	3	4	5	6	7	8	9	10	11	12	13	14	15	16
A	u_k		1	1	0	0	1	0	0	1	0	0	1	0	0	1	1	0
B	E1		3	2	0	1	2	1	0	2	0	1	2	1	0	2	3	0
C	$u_{I[k]}$		1	1	0	1	1	0	0	0	1	0	0	0	1	1	0	0
D	E2		3	2	0	2	3	0	0	0	3	1	1	0	2	3	0	0
E	$x_k^{(1s)}$		1	1	−1	−1	1	−1	−1	1	−1	−1	1	−1	−1	1	1	−1
F	$y_k^{(1s)}$		0.692	1.469	−1.593	−0.331	0.309	−0.342	−1.573	1.442	−1.275	−0.913	1.109	−1.295	−0.542	0.416	1.663	−1.689
G	$x_k^{(1p)}$		1	−1	−1	1	−1	1	−1	−1	−1	1	−1	1	−1	−1	1	−1
H	$y_k^{(1p)}$		1.109	0	−0.542	0	−0.337	0	−0.34	0	−0.549	0	−0.9	0	−1.282	0	0.425	0
I	$x_k^{(2p)}$		1	−1	−1	−1	1	−1	−1	−1	1	1	1	−1	−1	1	−1	−1
J	$y_k^{(2p)}$		0	−0.905	0	−0.554	0	−0.343	0	−0.338	0	1.459	0	−0.887	0	0.731	0	−1.566
K	$y_{I[k]}^{(1s)}$		0.692	0.416	−1.275	1.442	1.663	−0.913	−1.573	−0.331	1.109	−0.342	−1.593	−1.689	0.309	1.469	−0.542	−1.295
L	$\widetilde{\alpha}_k^{(1)}(0)$	1	0.0266	0.00134	0.0148	0.0159	0.1489	0.1526	0.1984	0.1854	0.2407	0.2174	0.0686	0.0738	0.1049	0.0765	0.686	0.6654
M	$\widetilde{\alpha}_k^{(1)}(1)$	0	0	0.9245	0.0182	0.3325	0.1601	0.4076	0.1847	0.4838	0.0727	0.1622	0.1433	0.6705	0.0641	0.5562	0.0608	0.1471
N	$\widetilde{\alpha}_k^{(1)}(2)$	0	0.9734	0.0252	0.9561	0.017	0.1566	0.1563	0.5046	0.1976	0.594	0.096	0.0728	0.1381	0.773	0.0926	0.1046	0.0814
O	$\widetilde{\alpha}_k^{(1)}(3)$	0	0	0.0489	0.0109	0.6346	0.5344	0.2834	0.1123	0.1331	0.0927	0.5245	0.7153	0.1177	0.058	0.2747	0.1486	0.1061
P	$\widetilde{\beta}_k^{(1)}(0)$	0.6654	0.193	0.3183	0.2982	0.2884	0.1206	0.143	0.0898	0.0999	0.0547	0.0513	0.0447	0.0451	0.0176	0.0183	0.967	1
Q	$\widetilde{\beta}_k^{(1)}(1)$	0.2059	0.3116	0.7074	0.3074	0.2989	0.0987	1.5023	0.0993	1.2871	0.0722	0.0395	0.0394	0.9545	0.018	1.1901	0.033	0
R	$\widetilde{\beta}_k^{(1)}(2)$	0.0715	0.6783	0.1864	0.6798	0.3172	0.5492	0.0763	1.2215	0.0892	1.0575	0.0756	0.1989	0.039	0.8297	0.0173	0	0
S	$\widetilde{\beta}_k^{(1)}(3)$	0.0928	0.1578	0.1287	0.4953	0.8761	1.021	0.0679	0.1142	0.0485	0.2034	1.2214	0.8978	0.1423	0.3618	0.00146	0	0
T	$L_{12}^{(e)}(u_k)$		3.474	1.646	−1.626	−1.025	1.097	−1.019	−1.201	1.214	−1.539	−1.55	2.092	−1.358	−2.405	2.535	2.585	−2.423
U	$L_{12}^{(e)}(u_{I[k]})$		3.474	2.535	−1.539	1.214	2.585	−1.55	−1.201	−1.025	2.092	−1.019	−1.626	−2.423	1.097	1.646	−2.405	−1.358
V	$\widetilde{\alpha}_k^{(2)}(0)$	1	0.0077	0.0003	0.0166	0.0021	0.9828	0.9716	0.9592	0.8892	0.0248	0.0183	0.0183	0.0194	0.0056	0.9446	0.9168	0.9551
W	$\widetilde{\alpha}_k^{(2)}(1)$	0	0	0.9929	0.0055	0.9855	0.0056	0.0035	0.0029	0.0131	0.0853	0.0226	0.9417	0.00313	0.8255	0.0063	0.0396	0.0011

	k	0	1	2	3	4	5	6	7	8	9	10	11	12	13	14	15	16
X	$\widetilde{\alpha}_k^{(2)}(2)$	0	0.9923	0.0013	0.9765	0.0056	0.0048	0.0223	0.0159	0.0863	0.8776	0.0098	0.0226	0.9734	0.0169	0.0087	0.0341	0.042
Y	$\widetilde{\alpha}_k^{(2)}(3)$	0	0	0.0056	0.0013	0.0068	0.0068	0.0026	0.0221	0.0113	0.0123	0.9492	0.0174	0.0040	0.152	0.0405	0.0096	0.0017
Z	$\widetilde{\beta}_k^{(2)}(0)$	0.914	0.0051	0.027	0.011	0.049	0.929	0.94	0.952	1.027	0.0023	0.0087	0.0085	0.0077	0.0042	0.966	0.995	0.9551
A*	$\widetilde{\beta}_k^{(2)}(1)$	0.012	0.0053	0.92	0.017	0.927	0.032	0.035	0.092	0.016	0.0022	0.031	0.97	0.0049	1.105	0.03	0.045	0.0011
B*	$\widetilde{\beta}_k^{(2)}(2)$	0.0054	0.921	0.026	0.936	0.032	0.047	0.016	0.023	0.0028	1.041	0.0092	0.031	0.938	0.0083	0.0014	0.0001	0.042
C*	$\widetilde{\beta}_k^{(2)}(3)$	0.022	0.022	0.017	0.026	0.021	0.021	0.091	0.015	0.043	0.044	0.962	0.0014	0.179	0.012	0.043	0.00009	0.0017
D*	$L_{21}^{(e)}\left(u_{D[k]}\right)$		5.189	4.773	−4.369	−5.488	4.205	−5.614	−3.498	4.723	−4.369	−4.396	3.858	−5.557	−4.942	5.729	2.944	−4.787
E*	$L^{(1)}(u_k)$		10.046	9.358	−9.181	−7.176	5.919	−7.315	−7.844	8.82	−8.459	−7.773	8.167	−9.506	−8.432	9.096	8.855	−10.589
F*	$\widehat{u_k}$		1	1	0	0	1	0	0	1	0	0	1	0	0	1	1	0

To simulate the use of BPSK over an AWGN channel, a Gaussian random number generator is used with the mean set to +1 or -1 if the code digit to be transmitted is either a binary digit 1 or 0, respectively with a standard deviation $\sigma = \sqrt{\frac{0.5}{10^{SNR/10}}}$. Row E lists the first systematic code digit $x_k^{(1s)}$ of the codeword in row B and the corresponding received BPSK signal points $y_k^{(1s)}$ are listed in row F for the $SNR =$ 0 dB.

Similarly, row G lists the second code digit (parity bit of E1) of the codeword in row B and row I lists the second code digit (parity bit of E2) of the codeword in row D. The corresponding received signal points are listed in rows H and J with alternating values punctured to 0. For example, in row H, the $y_k^{(1p)}$ values at time index 2, 4, 6, 8, 10, 12, 14, and 16 are set to 0. The turbo decoder thus receives the pairs $\left(y_1^{(1s)}, y_1^{(1p)}\right)$, $\left(y_2^{(1s)}, y_2^{(2p)}\right)$, $\left(y_3^{(1s)}, y_3^{(1p)}\right)$, $\left(y_4^{(1s)}, y_4^{(2p)}\right)$, $\left(y_5^{(1s)}, y_5^{(1p)}\right)$, $\left(y_6^{(1s)}, y_6^{(2p)}\right)$, $\left(y_7^{(1s)}, y_7^{(1p)}\right)$, $\left(y_8^{(1s)}, y_8^{(2p)}\right)$, $\left(y_9^{(1s)}, y_9^{(1p)}\right)$, $\left(y_{10}^{(1s)}, y_{10}^{(2p)}\right)$, $\left(y_{11}^{(1s)}, y_{11}^{(1p)}\right)$, $\left(y_{12}^{(1s)}, y_{12}^{(2p)}\right)$, $\left(y_{13}^{(1s)}, y_{13}^{(1p)}\right)$, $\left(y_{14}^{(1s)}, y_{14}^{(2p)}\right)$,$\left(y_{15}^{(1s)}, y_{15}^{(1p)}\right)$,$\left(y_{16}^{(1s)}, y_{16}^{(2p)}\right)$. Finally, given that $x_k^{(2s)} = u_{I[k]}$, the $y_k^{(2s)} = y_{I[k]}^{(1s)}$ are listed in row K by interleaving row F.

4.7.1 D1 Modified Probabilities

We shall now use the equations in Table 4.6 to determine the decoder D1 probabilities $\widetilde{\alpha}_k^{(1)}(s)$ at time index k of 0 and 1 only. Given that the encoder starts in the all-zero state, $\widetilde{\alpha}_0^1(0) = 1$ and the remaining $\widetilde{\alpha}_0^{(1)}(s)$ must be 0 to satisfy $\sum_s \widetilde{\alpha}_0^{(1)}(s) = 1$. To find the $\widetilde{\alpha}_1^{(1)}(s)$ values, we require $\gamma_1(s', s) = \gamma_1^{(1e)}(s', s) \exp\left[\frac{u_1}{2}\left(L^{(e)}(u_1) + L_c y_1^{(1s)}\right)\right]$ in which $L^{(e)}(u_1) = L_{21}^{(e)}(u_1) = 0$ because extrinsic information from D2 is not yet available.

Referring to the RSC code trellis shown in Fig. 4.19, there are two branches coming out of state $s = 0$. For the branch 0/00 corresponding to $\left(s' = 0\right) \longrightarrow (s = 0)$, the first code digit on this branch is "0" thus we set $u_1 = -1$. Then $\gamma_1^{(1e)}(0, 0) = \exp\left[\frac{L_c y_1^{(1p)} x_1^{(1p)}}{2}\right]$ in which $L_c = 4R_{code}\left(10^{SNR/10}\right) = 4(\frac{1}{2})\left(10^{0/10}\right) = 2$, $y_1^{(1p)} = 1.109$ on row H.

The second code digit on the branch 0/00 is "0," thus we set $x_1^{(1p)} = -1$. Note carefully that this is **not** the transmitted parity bit from E1 but the parity bit on the branch $\left(s' = 0\right) \longrightarrow (s = 0)$ under consideration. Therefore, $\gamma_1^{(1e)}(0, 0) =$

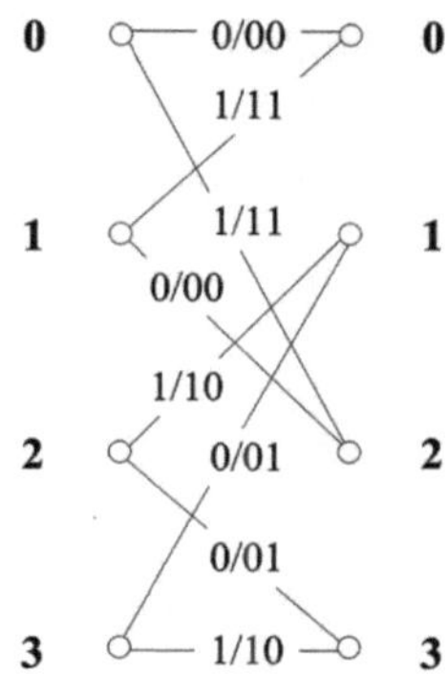

Fig. 4.19 Rate 1/2 RSC code trellis

$\exp\left[\frac{2(1.109)(-1)}{2}\right] = 0.33$ so that $\gamma_1(0,0) = 0.33\exp\left[\frac{-1}{2}\left(0+2y_1^{(1s)}\right)\right] = 0.33\exp\left[\frac{-1}{2}(2(0.692))\right] = 0.165$.

For the branch 1/11 corresponding to $\left(s'=0\right) \longrightarrow (s=2)$, the first code digit is "1" thus we set $u_1 = 1$. Then $\gamma_1^{(1e)}(0,2) = \exp\left[\frac{L_c y_1^{(1p)} x_1^{(1p)}}{2}\right]$ in which $L_c = 2$, $y_1^{(1p)} = 1.109$ and $x_1^{(1p)}$ is "1," thus we set $x_1^{(1p)} = 1$. Therefore, $\gamma_1^{(1e)}(0,2) = \exp\left[\frac{2(1.109)(1)}{2}\right] = 3.031$ and $\gamma_1(0,2) = 3.031\exp\left[\frac{1}{2}\left(0+2y_1^{(1s)}\right)\right] = 3.031\exp\left[\frac{1}{2}(2(0.692))\right] = 6.056$. Repeating this method and minimizing rounding errors, the $\gamma_1(s',s)$ for all possible transitions (s',s) within the code trellis at $k=0$ are

$$\begin{bmatrix} \gamma_1(0,0) & \gamma_1(0,1) & \gamma_1(0,2) & \gamma_1(0,3) \\ \gamma_1(1,0) & \gamma_1(1,1) & \gamma_1(1,2) & \gamma_1(1,3) \\ \gamma_1(2,0) & \gamma_1(2,1) & \gamma_1(2,2) & \gamma_1(2,3) \\ \gamma_1(3,0) & \gamma_1(3,1) & \gamma_1(3,2) & \gamma_1(3,3) \end{bmatrix} = \begin{bmatrix} 0.165 & 0 & 6.053 & 0 \\ 6.053 & 0 & 0.165 & 0 \\ 0 & 0.659 & 0 & 1.517 \\ 0 & 1.517 & 0 & 0.659 \end{bmatrix} \tag{4.28}$$

where for example, $\left(s'=3\right) \longrightarrow (s=1)$, $\gamma_1(3,1) = 1.517$. All the zero entries are of course transitions, which cannot take place within the code trellis. From Eq. 4.23, and taking into account that $\widetilde{\alpha}_0^{(1)}(0) = 1$ and $\widetilde{\alpha}_0^{(1)}(s \neq 0) = 0$, the modified probability $\widetilde{\alpha}_1^{(1)}(2)$ is given by

$$\begin{aligned} \widetilde{\alpha}_1^{(1)}(2) &= \frac{\sum_{s'} \widetilde{\alpha}_0^{(1)}(s')\gamma_1(s',2)}{\sum_{s}\sum_{s'} \widetilde{\alpha}_0^{(1)}(s')\gamma_1(s',s)} = \frac{\gamma_1(0,2)}{\gamma_1(0,0)+\gamma_1(0,2)} = \frac{6.053}{0.165+6.053} \\ &= 0.9735 \end{aligned} \tag{4.29}$$

Similarly, $\widetilde{\alpha}_1^{(1)}(0) = \frac{\gamma_1(0,0)}{\gamma_1(0,0)+\gamma_1(0,2)} = \frac{0.165}{0.165+6.053} = 0.0265$. These probabilities minimizing rounding errors are listed in rows L, M, N, and O of Table 4.7 where we find as expected, $\sum_s \widetilde{\alpha}_k^{(1)}(s) = 1$. Now from Eq. 4.24, for $k = N - 1 = 15$,

$$\widetilde{\beta}_{15}^{(1)}(s') = \frac{\sum_s \widetilde{\beta}_{16}^{(1)}(s)\gamma_{16}(s', s)}{\sum_s \sum_{s'} \widetilde{\alpha}_{15}^{(1)}(s')\gamma_{16}(s', s)} \tag{4.30}$$

where we find that

$$\begin{bmatrix} \gamma_{16}(0,0) & \gamma_{16}(0,1) & \gamma_{16}(0,2) & \gamma_{16}(0,3) \\ \gamma_{16}(1,0) & \gamma_{16}(1,1) & \gamma_{16}(1,2) & \gamma_{16}(1,3) \\ \gamma_{16}(2,0) & \gamma_{16}(2,1) & \gamma_{16}(2,2) & \gamma_{16}(2,3) \\ \gamma_{16}(3,0) & \gamma_{16}(3,1) & \gamma_{16}(3,2) & \gamma_{16}(3,3) \end{bmatrix} = \begin{bmatrix} 5.416 & 0 & 0.185 & 0 \\ 0.185 & 0 & 5.416 & 0 \\ 0 & 0.185 & 0 & 5.416 \\ 0 & 5.416 & 0 & 0.185 \end{bmatrix} \tag{4.31}$$

Given the initialization $\widetilde{\beta}_{16}^{(1)}(0) = 1$ and $\widetilde{\beta}_{16}^{(1)}(s \neq 0) = 0$,

$$\widetilde{\beta}_{15}^{(1)}(0) = \frac{\sum_s \widetilde{\beta}_{16}^{(1)}(s)\gamma_{16}(0, s)}{\sum_s \sum_{s'} \widetilde{\alpha}_{15}^{(1)}(s')\gamma_{16}(s', s)} \tag{4.32}$$

$$= \frac{\gamma_{16}(0,0)}{\sum_s \left[\widetilde{\alpha}_{15}^{(1)}(0)\gamma_{16}(0,s) + \widetilde{\alpha}_{15}^{(1)}(1)\gamma_{16}(1,s) + \widetilde{\alpha}_{15}^{(1)}(2)\gamma_{16}(2,s) + \widetilde{\alpha}_{15}^{(1)}(3)\gamma_{16}(3,s)\right]} \tag{4.33}$$

where for example, $\sum_s \widetilde{\alpha}_{15}^{(1)}(0)\gamma_{16}(0,s) = \widetilde{\alpha}_{15}^{(1)}(0)\,(\gamma_{16}(0,0) + \gamma_{16}(0,2))$. Given the symmetry of the $\gamma_{16}(.)$ values and $\sum_s \widetilde{\alpha}_{15}^{(1)}(s) = 1$, we find that for this time index only,

$$\widetilde{\beta}_{15}^{(1)}(0) = \frac{\gamma_{16}(0,0)}{\gamma_{16}(0,0) + \gamma_{16}(0,2)} = 0.967 \tag{4.34}$$

and similarly, $\widetilde{\beta}_{15}^{(1)}(1) = \frac{\gamma_{16}(0,2)}{\gamma_{16}(0,0)+\gamma_{16}(0,2)} = 0.033$. The same procedure is used to then determine the remaining modified probabilities listed in rows P, Q, R, and S. An interesting feature of these modified probabilities is that $\sum_s \widetilde{\alpha}_k^{(1)}(s)\widetilde{\beta}_k^{(1)}(s)$ is a constant value for any value of k [16]. For the probabilities listed in Table 4.7, this value is 0.6654, which corresponds to the value of $\widetilde{\beta}_0^{(1)}(0) = \widetilde{\alpha}_N^{(1)}(0)$. The next step is to determine the extrinsic information $L_{12}^{(e)}(u_k)$ to be passed from D1 to D2, which for example for $k = 1$ is given by

$$L_{12}^{(e)}(u_1) = \log\left(\frac{\sum_{S^+} \widetilde{\alpha}_0^{(1)}(s')\gamma_1^{(1e)}(s',s)\widetilde{\beta}_1^{(1)}(s)}{\sum_{S^-} \widetilde{\alpha}_0^{(1)}(s')\gamma_1^{(1e)}(s',s)\widetilde{\beta}_1^{(1)}(s)}\right) \tag{4.35}$$

The S^+ in the numerator are the set of state transitions $(s_0 = s') \longrightarrow (s_1 = s)$ that are the result of an input information digit "1" to the encoder at time $k = 1$. Thus, for the numerator and taking into account $\widetilde{\alpha}_0^{(1)}(s \neq 0) = 0$, we find $\widetilde{\alpha}_0^{(1)}(0)\gamma_1^{(1e)}(0,2)\widetilde{\beta}_1^{(1)}(2) + \widetilde{\alpha}_0^{(1)}(1)\gamma_1^{(1e)}(1,0)\widetilde{\beta}_1^{(1)}(0) + \widetilde{\alpha}_0^{(1)}(2)\gamma_1^{(1e)}(2,1)\widetilde{\beta}_1^{(1)}(1) + \widetilde{\alpha}_0^{(1)}(3)\gamma_1^{(1e)}(3,3)\widetilde{\beta}_1^{(1)}(3) = \widetilde{\alpha}_0^{(1)}(0)\gamma_1^{(1e)}(0,2)\widetilde{\beta}_1^{(1)}(2) = (1)(3.031)(0.6783) = 2.056$.

Similarly, the S^- in the denominator are the set of state transitions $(s_0 = s') \longrightarrow (s_1 = s)$ that are the result of an input information digit "0" to the encoder at time $k = 1$. Thus, for the denominator and taking into account $\widetilde{\alpha}_0^{(1)}(s \neq 0) = 0$, we find $\widetilde{\alpha}_0^{(1)}(0)\gamma_1^{(1e)}(0,0)\widetilde{\beta}_1^{(1)}(0) + \widetilde{\alpha}_0^{(1)}(1)\gamma_1^{(1e)}(1,2)\widetilde{\beta}_1^{(1)}(2) + \widetilde{\alpha}_0^{(1)}(2)\gamma_1^{(1e)}(2,3)\widetilde{\beta}_1^{(1)}(3) + \widetilde{\alpha}_0^{(1)}(3)\gamma_1^{(1e)}(3,1)\widetilde{\beta}_1^{(1)}(1) = \widetilde{\alpha}_0^{(1)}(0)\gamma_1^{(1e)}(0,0)\widetilde{\beta}_1^{(1)}(0) = (1)(0.33)(0.193) = 0.064$. Thus,

$$L_{12}^{(e)}(u_1) = \log\left(\frac{2.056}{0.064}\right) = 3.47 \tag{4.36}$$

4.7.2 D2 Modified Probabilities

All the values of $L_{12}^{(e)}(u_k)$ are listed in row T are then interleaved to produce $L_{12}^{(e)}(u_{I[k]})$ in row U. Using $L_{12}^{(e)}(u_{I[k]})$ and $\left(y_{I[k]}^{(1s)}, y_k^{(2p)}\right)$, the $\gamma_k(s',s)$ probabilities for D2 are calculated. For example, for $k = 1$,

$$\gamma_1(s',s) = \gamma_1^{(2e)}(s',s)\exp\left[\frac{u_1}{2}\left(L^{(e)}(u_1) + L_c y_1^{(2s)}\right)\right] \tag{4.37}$$

in which $L^{(e)}(u_1) = L_{12}^{(e)}(u_{I[1]}) = 3.474$ from row U, $y_1^{(2s)} = y_{I[1]}^{(1s)} = 0.692$ from row K and

$$\gamma_1^{(2e)}(s',s) = \exp\left[\frac{L_c y_1^{(2p)} x_1^{(2p)}}{2}\right] \tag{4.38}$$

For the branch 0/00 corresponding to $(s' = 0) \longrightarrow (s = 0)$, the second code digit (parity bit) on this branch is '0' thus we set $u_1 = -1$. Given that $y_1^{(2p)} = 0$, $\gamma_1^{(2e)}(0,0) = 1$ and therefore $\gamma_1(0,0) = (1)\exp\left[\frac{-1}{2}(3.474 + 2(0.692))\right] = 0.088$.

Similarly for the branch 1/11 corresponding to $\left(s' = 0\right) \longrightarrow (s = 2)$, the second code digit (parity bit) on this branch is "1"; thus, we set $u_1 = 1$. Given that $y_1^{(2p)} = 0$, $\gamma_1^{(2e)}(0, 2) = 1$ and therefore $\gamma_1(0, 2) = (1)\exp\left[\frac{1}{2}(3.474 + 2(0.692))\right] = 11.348$.

Following the same procedure as in D1 using Eq. 4.23, and taking into account that $\widetilde{\alpha}_0^{(2)}(0) = 1$ and $\widetilde{\alpha}_0^{(2)}(s \neq 0) = 0$, the modified probability $\widetilde{\alpha}_1^{(2)}(2)$ is given by

$$\widetilde{\alpha}_1^{(2)}(2) = \frac{\sum\limits_{s'} \widetilde{\alpha}_0^{(2)}(s')\gamma_1(s', 2)}{\sum\limits_{s}\sum\limits_{s'} \widetilde{\alpha}_0^{(2)}(s')\gamma_1(s', s)} = \frac{\gamma_1(0, 2)}{\gamma_1(0, 0) + \gamma_1(0, 2)} = \frac{11.348}{0.088 + 11.348}$$
$$= 0.992 \tag{4.39}$$

which we note has increased in comparison to $\widetilde{\alpha}_1^{(1)}(2) = 0.9734$ because of the extrinsic information. All the values of $\widetilde{\alpha}_k^{(2)}(s)$ are listed in rows V–Y. An **important** next step is to set $\widetilde{\alpha}_N^{(2)}(s) = \widetilde{\beta}_N^{(2)}(s)$ as shown in Table 4.7 rows Z to C* at time index $k = 16$. Thereafter, the $\widetilde{\beta}_k^{(2)}(s)$ values are calculated backward from $k = 16$ to $k = 1$, and we find that $\sum\limits_{s} \widetilde{\alpha}_k^{(2)}(s)\widetilde{\beta}_k^{(2)}(s) = 0.914$. The next step is to determine the extrinsic information $L_{21}^{(e)}(u_k)$ to be passed from D2 to D1, which for $k = 1$ is given by

$$L_{21}^{(e)}(u_1) = \log\left(\frac{\sum\limits_{S^+} \widetilde{\alpha}_0^{(2)}(s')\gamma_1^{(2e)}(s', s)\widetilde{\beta}_1^{(2)}(s)}{\sum\limits_{S^-} \widetilde{\alpha}_0^{(2)}(s')\gamma_1^{(2e)}(s', s)\widetilde{\beta}_1^{(2)}(s)}\right) \tag{4.40}$$

For the numerator given $\widetilde{\alpha}_0^{(2)}(s \neq 0) = 0$, we find $\widetilde{\alpha}_0^{(2)}(0)\gamma_1^{(2e)}(0, 2)\widetilde{\beta}_1^{(2)}(2) + \widetilde{\alpha}_0^{(2)}(1)\gamma_1^{(2e)}(1, 0)\widetilde{\beta}_1^{(2)}(0) + \widetilde{\alpha}_0^{(2)}(2)\gamma_1^{(2e)}(2, 1)\widetilde{\beta}_1^{(2)}(1) + \widetilde{\alpha}_0^{(2)}(3)\gamma_1^{(2e)}(3, 3)\widetilde{\beta}_1^{(2)}(3) = \widetilde{\alpha}_0^{(2)}(0)\gamma_1^{(2e)}(0, 2)\widetilde{\beta}_1^{(2)}(2) = (1)(1)(0.921) = 0.921$.

Similarly, for the denominator and taking into account $\widetilde{\alpha}_0^{(2)}(s \neq 0) = 0$, we find $\widetilde{\alpha}_0^{(2)}(0)\gamma_1^{(2e)}(0, 0)\widetilde{\beta}_1^{(2)}(0) + \widetilde{\alpha}_0^{(2)}(1)\gamma_1^{(2e)}(1, 2)\widetilde{\beta}_1^{(2)}(2) + \widetilde{\alpha}_0^{(2)}(2)\gamma_1^{(2e)}(2, 3)\widetilde{\beta}_1^{(2)}(3) + \widetilde{\alpha}_0^{(2)}(3)\gamma_1^{(2e)}(3, 1)\widetilde{\beta}_1^{(2)}(1) = \widetilde{\alpha}_0^{(2)}(0)\gamma_1^{(2e)}(0, 0)\widetilde{\beta}_1^{(2)}(0) = (1)(1)(0.00514) = 0.00514$. Thus

$$L_{21}^{(e)}(u_1) = \log\left(\frac{0.921}{0.00514}\right) = 5.189. \tag{4.41}$$

This extrinsic information is then **deinterleaved** to create $L_{21}^{(e)}\left(u_{D[k]}\right)$ as shown in row D*. For the pseudo-interleaving used in this example, position 1 does not move, which is why the value 5.188 is in row D* at time index $k = 1$ after deinterleaving. The final step is to determine the log-likelihood ratio $L^{(1)}(u_1)$ for

Simulation 4.9: TurboDecoderSteps

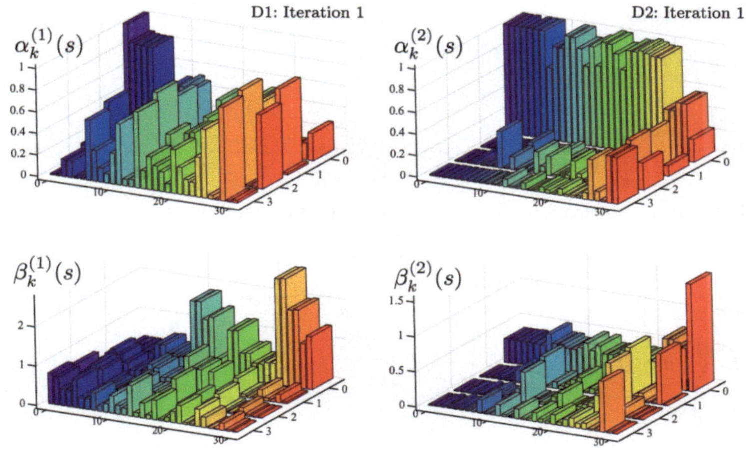

Fig. 4.20 $\widetilde{\alpha}_k(s)$ and $\widetilde{\beta}_k(s)$ probabilities within D1 and D2 for $k = 0, 1, \ldots, 30$ only

D1, which at $k = 1$ is given by

$$L^{(1)}(u_1) = L_c y_1^{(1s)} + L_{21}^{(e)}\left(u_{D[1]}\right) + L_{12}^{(e)}(u_1) = 2(0.692) + 5.189 + 3.474 = 10.047. \tag{4.42}$$

All the values of $L^{(1)}(u_k)$ are shown in row E* minimizing rounding errors. At time k, the estimate $\widehat{u_k} = 1$ if $L^{(1)}(u_k) > 0$; otherwise, $\widehat{u_k} = 0$. Accordingly, the decoded information bits are shown in row F* to complete the first iteration of the turbo decoder. In this simple example, the decoded information is the same as the transmitted information and no further iterations of the decoder are required.

However, for $M_I = 32$, Fig. 4.20 illustrates the typical $\widetilde{\alpha}_k(s)$ and $\widetilde{\beta}_k(s)$ values within D1 and D2, respectively, after the first iteration if the information sequence $\{u_k\}$ is an all-zero sequence. Only the values for $k = 0, 1, \ldots 30$ are shown in this figure. Notice how the modified probabilities $\widetilde{\alpha}_k^{(2)}(0)$ within D2 are improved because an all-zero sequence was sent. The corresponding $L_{12}^{(e)}(u_k)$, $L_{21}^{(e)}\left(u_{D[k]}\right)$ and $L_c y_k^{(1s)}$ values for the first three iterations are presented in Figs. 4.21, 4.22, and 4.23. Notice how the extrinsic information increases as the decoder iterates toward a decoding solution.

4.8 Simulation: Turbo Decoder

In simulation TurboDecoder, the DCS shown in Fig. 4.24 is modeled. The corresponding simulation results are presented in Fig. 4.25 for the probability of

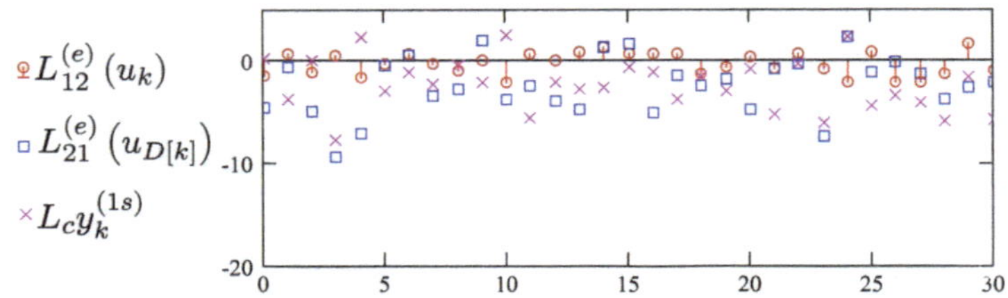

Fig. 4.21 Extrinsic information and channel value after first iteration

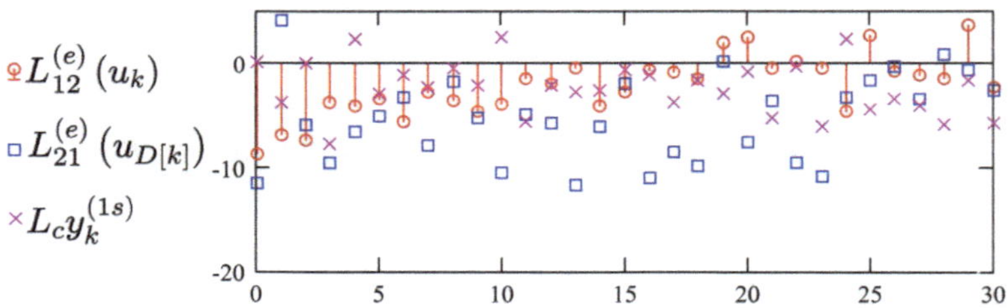

Fig. 4.22 Extrinsic information and channel value after second iteration

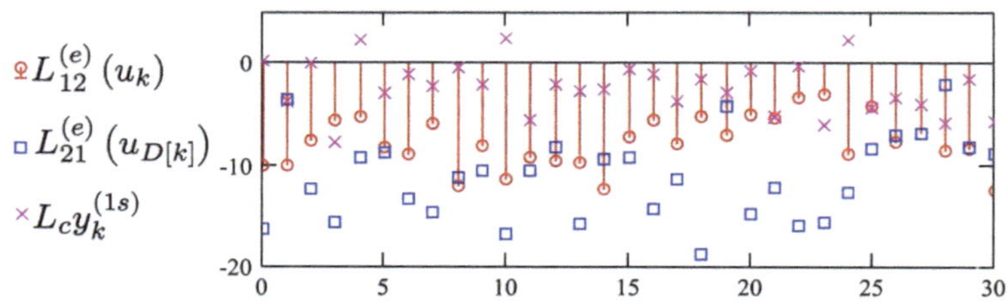

Fig. 4.23 Extrinsic information and channel value after third iteration

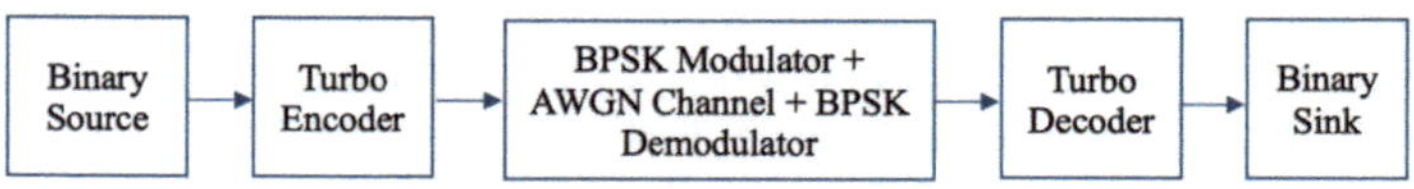

Fig. 4.24 Simulated DCS : BPSK over an AWGN channel with Turbo decoding

an error in a binary digit P_e at the binary sink. We shall use the notation $R_{code}(G_1, G_2, M_I * M_I)$ to represent a turbo code encoder with code rate R_{code}, that consists of two identical RSC code encoders (E1 = E2) with generator polynomials G_1 and G_2, respectively and using a square interleaver of size M_I x M_I.

For $M_I = 32$, the binary digits output from the binary source are processed in blocks of 1024 digits. Notice how the performance of the turbo decoder in Fig. 4.25 improves with each iteration initially. The results indicate that for this code, there is little to be gained by allowing the decoder to iterate beyond four iterations. In [14],

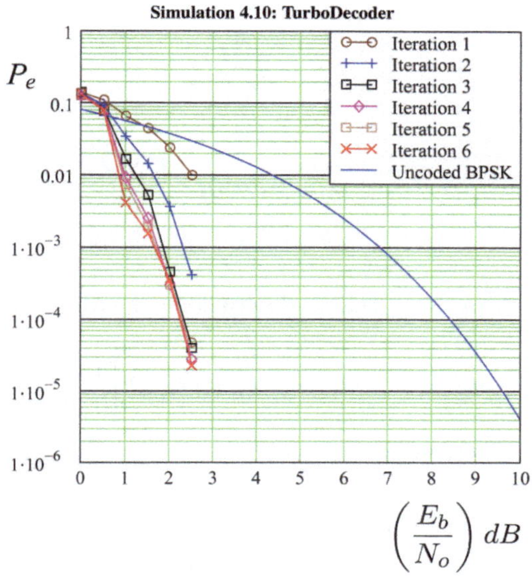

Fig. 4.25 Performance of the rate $1/2\,(7, 5, 32 * 32)$ encoder using BPSK over an AWGN

the rate $1/2$ $(37, 21, 256 * 256)$ provides $P_e = 10^{-5}$ at a SNR of 0.7 dB using BPSK over an AWGN channel.

Experiment Repeat the simulation for $M_I = 64$. Change the parameters of the interleaver or try another type of interleaver. If you are undertaking research in this area, get your feet wet by implementing the rate $1/2$ $(37, 21, 256 * 256)$ turbo code [14].

4.9 Problems

1. Consider a rate $\frac{1}{2}$, constraint length 3 convolutional code with a generator transfer function matrix $G(D) = \begin{bmatrix} 1 + D + D^2 & 1 + D^2 \end{bmatrix}$. If the code digit sequence received by the Viterbi decoder is $11, 00, 00, 10, 11, 00, 00$, where 11 is the first code word to be received, determine the following by assuming that the encoder started in the all-zero state.
   (a) The decoded information binary sequence.
   (b) Identify the code digits that were corrupted within the channel.
2. If the output of the encoder from the previous question is fed directly into the *syndrome former* shown below, what is the output sequence?

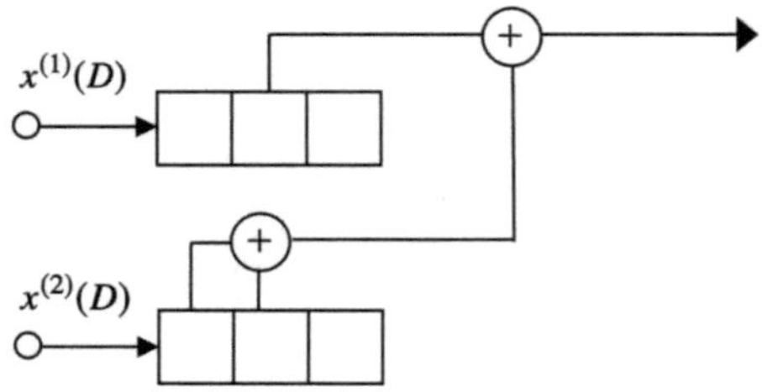

3. A rate $\frac{2}{3}$ code is described by the following generator matrix:

$$G(D) = \begin{bmatrix} 1+D & 1+D & 1 \\ D & 0 & 1+D \end{bmatrix}$$

   (a) Draw a diagram of the convolutional code encoder and determine its state diagram.
   (b) What is the free distance of this code?

4. A rate $\frac{1}{3}$ code is described by the following generator matrix:

$$G(D) = \begin{bmatrix} 1+D & 1+D+D^2 & 1+D^2 \end{bmatrix}$$

   (a) Draw a diagram of the convolutional code encoder and determine its state table.
   (b) Determine the free distance of this code.
   (c) What would be a near-optimum Viterbi trellis depth for this code in practice?

5. A rate $\frac{1}{2}$ convolutional code is described by the following generator matrix:

$$G(D) = [1+D \quad 1]$$

   (a) If the input sequence is 101111..., determine the corresponding output codeword sequence using the multiplication of polynomials only.
   (b) Draw a diagram of the convolutional code encoder and determine its state table. Use the state table to draw its code trellis diagram.
   (c) If the output sequence from the encoder in part (a) enters a channel, which corrupts every fourth bit, starting from the first bit that enters the channel, determine the output of the Viterbi decoder. For example, if the input to the channel is an all-zero sequence, the output of the channel is 10001000100010001...

6. A DCS consists of a binary source, a rate $\frac{1}{2}$ convolutional code encoder, a BSC channel, an MAP decoder, and a binary sink. The code trellis of the encoder is shown below.

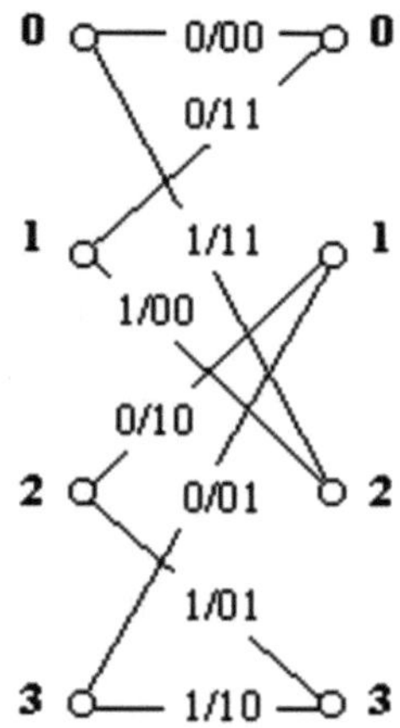

If the BSC crossover probability is 0, determine the following if the output of the binary source is the sequence 1011, where the digit 0 is the second digit output

(a) The transmitted code word sequence assuming that the encoder is flushed after encoding the information sequence 1011.
(b) The $\alpha_k(s)$ probabilities.
(c) The $\beta_k(s)$probabilities.
(d) Hence, find the $\lambda_k(s)$probabilities and decode the information sequence.

7. Determine the state table for the rate $\frac{1}{2}$, RSC encoder shown below.

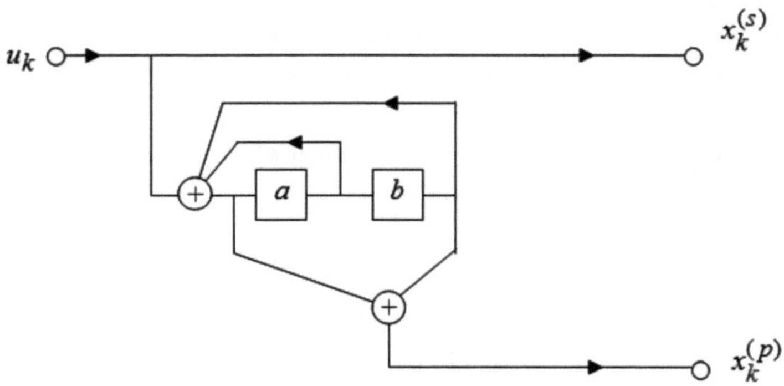

8. Determine the state table for the rate $\frac{1}{2}$, RSC encoder shown below.

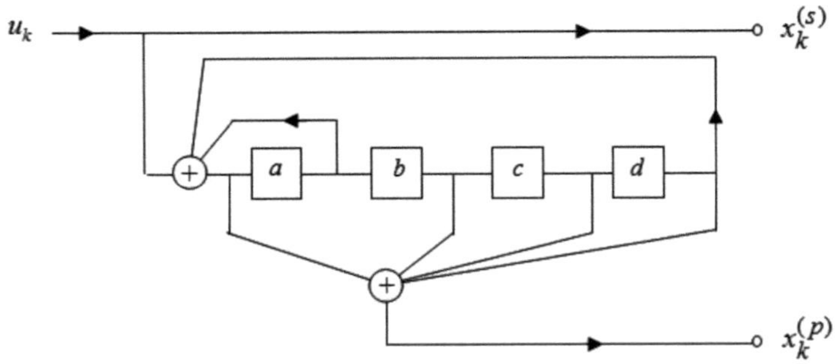

9. Determine the state table and code trellis diagram for the rate $\frac{1}{2}$, RSC encoder shown below.

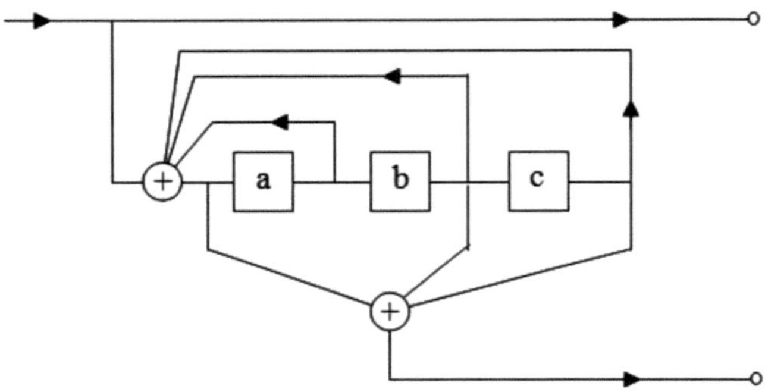

10. Show that $\gamma_k(s', s) = P\left(s_k = s|s_{k-1} = s'\right) P\left(y_k|b_k\right)$, where b_k is the branch codeword on the code trellis diagram between states s' and s.
11. Show that the probability $\lambda_k(s) = P(s_k = s, Y_1^N) = \alpha_k(s)\beta_k(s)$.
12. Show that $\sum_s \widetilde{\alpha}_k(s) = 1$.
13. Interleave the matrix shown below in which $M_I = 8$, using $P = \left[17\ 37\ 19\ 29\ 41\ 23\ 13\ 7\right]$.

0	1	2	3	4	5	6	7
8	9	10	11	12	13	14	15
16	17	18	19	20	21	22	23
24	25	26	27	28	29	30	31
32	33	34	35	36	37	38	39
40	41	42	43	44	45	46	47
48	49	50	51	52	53	54	55
56	57	58	59	60	61	62	63

14. A DCS consists of a binary source, a rate $\frac{1}{2}$ turbo code encoder, an AWGN channel, a turbo decoder, and a binary sink. The turbo code encoder consists of two identical rate $\frac{1}{2}$ RSC codes. A diagram of the overall encoder and the rate $\frac{1}{2}$ RSC code trellis is as follows, where E1 and E2 are the two encoders. The two corresponding modified MAP decoders will be referred to as D1 and D2. The channel signal-to-noise ratio is 2 dB. The interleaver matrix is of size (4×4).

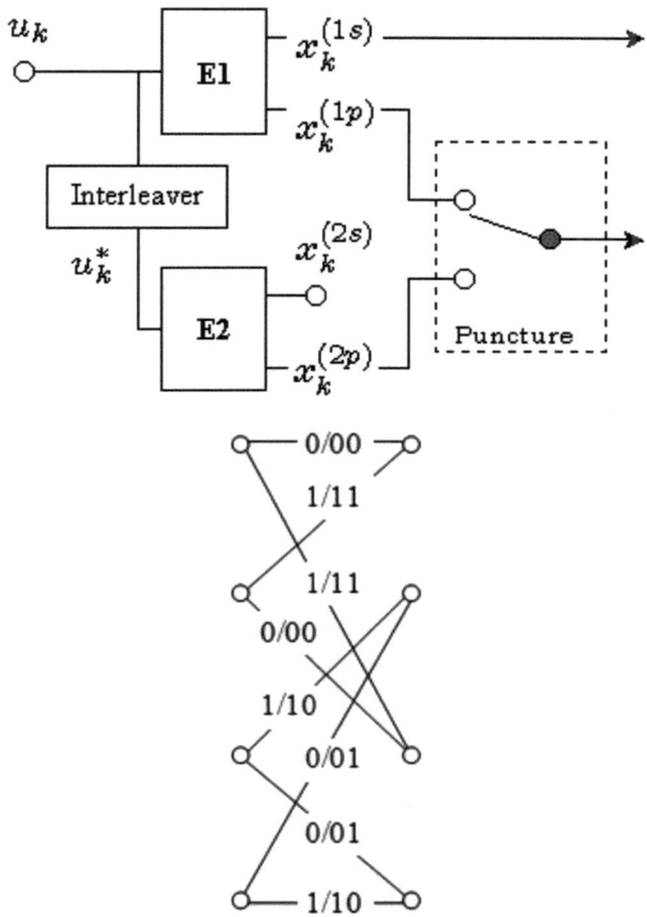

Given that the received code word sequence for D1 is received without error as follows, determine the $\gamma_k(s', s)$ values for $k = 1$ and $k = 15$.

15. Continuing the previous question, a sample of the modified probabilities $\tilde{\alpha}_k(s)$ within D1 as shown below. Verify the $\tilde{\alpha}_1(s)$ values.

k	$y_k^{(1s)}$	$y_k^{(1p)}$		k	$y_k^{(1s)}$	$y_k^{(1p)}$
1	−1	−1		9	−1	1
2	−1	0		10	−1	0
3	1	1		11	1	−1
4	1	0		12	1	0
5	−1	−1		13	−1	−1
6	−1	0		14	−1	0
7	1	−1		15	−1	−1
8	1	0		16	−1	0

	0	1	2	3	4	5	6
0	1	0.998	0.958	$1.829 \cdot 10^{-3}$	$1.829 \cdot 10^{-3}$	$3.654 \cdot 10^{-3}$	$3.583 \cdot 10^{-3}$
1	0	0	$7.104 \cdot 10^{-5}$	$1.829 \cdot 10^{-3}$	0.954	$1.908 \cdot 10^{-3}$	0.042
2	0	$1.762 \cdot 10^{-3}$	0.04	0.995	$1.829 \cdot 10^{-3}$	0.993	$1.979 \cdot 10^{-3}$
3	0	0	$1.691 \cdot 10^{-3}$	$1.829 \cdot 10^{-3}$	0.042	$1.908 \cdot 10^{-3}$	0.953

16. Continuing the previous question, verify the $\tilde{\beta}_{15}(s)$ values shown below.

	10	11	12	13	14	15	16
0	$2 \cdot 10^{-3}$	0.042	1	0.96	1	0.96	1
1	$2 \cdot 10^{-3}$	0.96	$3.604 \cdot 10^{-3}$	0.042	$1.765 \cdot 10^{-3}$	0.04	0
2	1	$3.536 \cdot 10^{-3}$	$1.916 \cdot 10^{-3}$	$1.765 \cdot 10^{-3}$	$1.765 \cdot 10^{-3}$	0	0
3	$3.832 \cdot 10^{-3}$	$1.984 \cdot 10^{-3}$	$1.916 \cdot 10^{-3}$	$1.765 \cdot 10^{-3}$	$1.765 \cdot 10^{-3}$	0	0

$\tilde{\beta}_k(s)$ from $k = 10$ to 16.

You may make use of the following $\tilde{\alpha}_k(s)$ values as required.

	10	11	12	13	14	15	16
0	0.042	$2.002 \cdot 10^{-3}$	0.952	0.992	0.952	0.992	0.952
1	$3.822 \cdot 10^{-3}$	0.992	$2.079 \cdot 10^{-3}$	$2.003 \cdot 10^{-3}$	$2.08 \cdot 10^{-3}$	$2.003 \cdot 10^{-3}$	$2.08 \cdot 10^{-3}$
2	0.952	$2.002 \cdot 10^{-3}$	0.042	$3.917 \cdot 10^{-3}$	0.042	$3.918 \cdot 10^{-3}$	0.042
3	$2.075 \cdot 10^{-3}$	$3.913 \cdot 10^{-3}$	$3.836 \cdot 10^{-3}$	$2.003 \cdot 10^{-3}$	$3.84 \cdot 10^{-3}$	$2.003 \cdot 10^{-3}$	$3.841 \cdot 10^{-3}$

17. Continuing the previous question, the extrinsic information $L_{12}^{(e)}(u_k)$ values passed from D1 to D2 is shown below. Verify the first value of –9.296 and decode the information sequence, without the use of $L_{21}^{(e)}\left(u_{D[k]}\right)$.
You may make use of the following $\tilde{\beta}_k(s)$ values as required.

	0	1	2	3	4	5	6
0	0.952	0.954	0.994	$2.091 \cdot 10^{-3}$	$3.938 \cdot 10^{-3}$	$2.094 \cdot 10^{-3}$	$2.017 \cdot 10^{-3}$
1	$3.761 \cdot 10^{-3}$	0.042	$3.925 \cdot 10^{-3}$	$3.861 \cdot 10^{-3}$	0.997	$3.863 \cdot 10^{-3}$	$2.017 \cdot 10^{-3}$
2	$1.923 \cdot 10^{-3}$	$2.084 \cdot 10^{-3}$	$2.007 \cdot 10^{-3}$	0.957	$2.014 \cdot 10^{-3}$	0.959	$3.941 \cdot 10^{-3}$
3	$1.923 \cdot 10^{-3}$	$3.848 \cdot 10^{-3}$	$2.007 \cdot 10^{-3}$	0.042	$2.014 \cdot 10^{-3}$	0.042	0.999

$\tilde{\beta}_k(s)$ from $k = 0$ to 6.

k	$L_{12}^{(e)}(u_k)$	k	$L_{12}^{(e)}(u_k)$	k	$L_{12}^{(e)}(u_k)$
1	−9.296	7	6.153	13	−6.164
2	−6.202	8	5.512	14	−5.574
3	6.239	9	−6.15	15	−6.244
4	5.555	10	−5.513	16	−6.205
5	−6.161	11	6.153		
6	−5.532	12	5.533		

18. Continuing the previous question, determine the value of $\sum_s \tilde{\alpha}_k(s)\tilde{\beta}_k(s)$ within D1 for any three different values of k. Make use of the $\tilde{\alpha}_k(s)$ and $\widetilde{\beta}_k(s)$ values given in the previous questions as required. What do you conclude?

Appendix A
Solutions to Chapter 1

[1] The two BSCs concatenated are shown below.

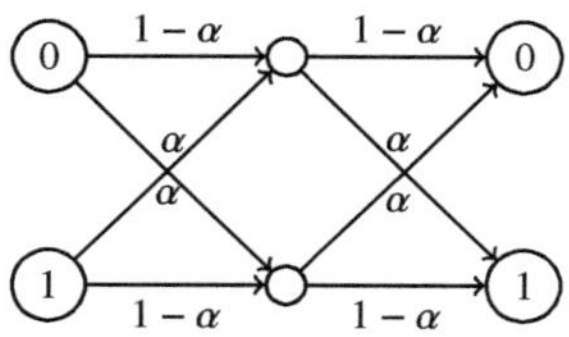

$$P(0 \mid 1) = \alpha(1-\alpha) + (1-\alpha)(\alpha) = 2\alpha(1-\alpha)$$

$$P(1 \mid 0) = (1-\alpha)(\alpha) + \alpha(1-\alpha) = 2\alpha(1-\alpha)$$

Hence equivalent BSC crossover probability $\boxed{\beta = 2\alpha(1-\alpha)}$

[2]

(i) The binary digits 0 and 1 are represented by

$$s_1(t) = A\cos(2\pi f_c t), \quad s_2(t) = -A\cos(2\pi f_c t),$$

where A is the amplitude and f_c is the carrier frequency. The energy of a BPSK signal is given by

$$E_b = \int_0^{T_b} s(t)^2\,dt$$

J. Sodha, *Fundamentals of Polar, LDPC and Turbo Codes*,
https://doi.org/10.1007/978-3-031-90924-5

For $s_1(t)$

$$E_b = \int_0^{T_b} A^2 \cos^2(2\pi f_c t)\, dt$$

We know that the average value of $\cos^2(\theta)$ over a period is $\frac{1}{2}$, namely

$$\frac{1}{T_b} \int_0^{T_b} \cos^2(2\pi f_c t)\, dt = \frac{1}{2}$$

$$\int_0^{T_b} \cos^2(2\pi f_c t)\, dt = \frac{T_b}{2}$$

$$\Rightarrow E_b = A^2 \cdot \frac{T_b}{2} = \boxed{\frac{A^2 T_b}{2}}$$

(ii) Using the basis function

$$\phi_1(t) = \sqrt{\frac{2}{T_b}} \cos(2\pi f_c t)$$

We can express A in terms of E_b

$$A = \sqrt{\frac{2E_b}{T_b}}$$

Then

$$s_1(t) = A\cos(2\pi f_c t) = \sqrt{\frac{2E_b}{T_b}} \cos(2\pi f_c t) = \sqrt{E_b} \cdot \phi_1(t)$$

$$s_2(t) = -\sqrt{E_b} \cdot \phi_1(t)$$

The signal point coordinate is

$$x_n = \int_0^{T_b} r(t)\phi_1(t)\, dt$$

For $s_1(t)$

$$x_n = \int_0^{T_b} \sqrt{E_b}\phi_1(t) \cdot \phi_1(t)\, dt = \sqrt{E_b} \int_0^{T_b} \phi_1^2(t)\, dt = \boxed{\sqrt{E_b}}$$

because

$$\int_0^{T_b} \phi_1^2(t)\, dt = 1 \quad \text{(unit energy basis function)}$$

Similarly, for $s_2(t)$

$$\boxed{x_n = -\sqrt{E_b}}$$

[3] For uncoded BPSK with hard-decision decoding, the conditional probability

$$P(0|1) = \int_0^{\infty} f(x \mid -\sqrt{E_b})\, dx$$

where the conditional PDF is

$$f(x \mid -\sqrt{E_b}) = \frac{1}{\sigma\sqrt{2\pi}} \exp\left[-\frac{(x+\sqrt{E_b})^2}{2\sigma^2}\right]$$

Now, let us make the substitution

$$-\frac{z^2}{2} = -\frac{(x+\sqrt{E_b})^2}{2\sigma^2} \quad \Rightarrow \quad z = \frac{x+\sqrt{E_b}}{\sigma}$$

Differentiating

$$\frac{dz}{dx} = \frac{1}{\sigma} \quad \Rightarrow \quad dx = \sigma\, dz$$

The integration limits change accordingly

$$x = 0 \Rightarrow z = \frac{\sqrt{E_b}}{\sigma}, \quad x = \infty \Rightarrow z = \infty$$

Substituting into the original integral

$$P(0|1) = \int_{\sqrt{E_b}/\sigma}^{\infty} \frac{1}{\sigma\sqrt{2\pi}} \exp\left(-\frac{z^2}{2}\right) \sigma\, dz$$

$$= \int_{\sqrt{E_b}/\sigma}^{\infty} \frac{1}{\sqrt{2\pi}} e^{-z^2/2} dz = Q\left(\frac{\sqrt{E_b}}{\sigma}\right)$$

Now, for coherent demodulation, the noise variance is

$$\sigma^2 = \frac{N_0}{2} \quad \Rightarrow \quad \sigma = \sqrt{\frac{N_0}{2}}$$

Substituting

$$\frac{\sqrt{E_b}}{\sigma} = \sqrt{\frac{2E_b}{N_0}}$$

Hence,

$$\boxed{P(0|1) = Q\left(\sqrt{\frac{2E_b}{N_0}}\right)}$$

[4] For an M-ary PSK system, the symbol duration is given by

$$T_s = T_b \log_2 M$$

For QPSK ($M = 4$)

$$T_s = T_b \log_2 4 = 2T_b$$

Given that the amplitude A and carrier frequency f_c are the same for both BPSK and QPSK, the $E_{\text{QPSK}} = 2E_b$ because its duration is twice that of BPSK.
[5] The basis function is given by

$$\phi_1(t) = \sqrt{\frac{2}{T_b}} \cos(2\pi f_c t + \theta)$$

To determine the BSC crossover probability α, we can make use of

$$\int_0^{T_b} \cos(2\pi f_c t) \cos(2\pi f_c t + \theta)\, dt = \frac{T_b \cos\theta}{2}$$

For $r(t) = s_1(t)$, the signal point coordinate

$$x_h = \int_0^{T_b} s_1(t)\phi_1(t)\, dt = \int_0^{T_b} \sqrt{\frac{2E_b}{T_b}} \cos(2\pi f_c t) \cdot \sqrt{\frac{2}{T_b}} \cos(2\pi f_c t + \theta)\, dt$$

$$= \frac{2\sqrt{E_b}}{T_b} \int_0^{T_b} \cos(2\pi f_c t) \cos(2\pi f_c t + \theta)\, dt$$

Using the given identity

$$x_h = \frac{2\sqrt{E_b}}{T_b} \cdot \frac{T_b \cos\theta}{2} = \sqrt{E_b}\cos\theta$$

Hence, the effective signal point has moved closer to 0. The BSC crossover probability is the Q-function evaluated at the new mean:

$$\alpha = Q\left(\frac{\sqrt{E_b}\cos\theta}{\sigma}\right)$$

With coherent demodulation, the noise standard deviation is:

$$\sigma = \sqrt{\frac{N_0}{2}} \quad \Rightarrow \quad \alpha = Q\left(\sqrt{\frac{2E_b}{N_0}}\cos\theta\right)$$

To find the value of θ such that $\alpha = \frac{1}{2}$:

$$Q(u) = \frac{1}{2} \Rightarrow u = 0 \Rightarrow \sqrt{\frac{2E_b}{N_0}}\cos\theta = 0$$

$$\Rightarrow \cos\theta = 0 \Rightarrow \theta = 90^\circ$$

$$\boxed{\theta = 90^\circ \text{ makes } \alpha = \frac{1}{2}}$$

[6]

(i) For $i = 1$,

$$s_1(t) = \sqrt{\frac{2E_s}{T_s}}\cos\left(2\pi f_c t + \frac{\pi}{4}\right)$$

Use the identity

$$\cos(a+b) = \cos a \cos b - \sin a \sin b$$

$$\Rightarrow s_1(t) = \sqrt{\frac{2E_s}{T_s}}\left[\cos(2\pi f_c t)\cos\left(\frac{\pi}{4}\right) - \sin(2\pi f_c t)\sin\left(\frac{\pi}{4}\right)\right]$$

$$= \sqrt{\frac{2E_s}{T_s}} \cdot \frac{1}{\sqrt{2}} [\cos(2\pi f_c t) - \sin(2\pi f_c t)]$$

$$= \sqrt{\frac{E_s}{T_s}} [\cos(2\pi f_c t) - \sin(2\pi f_c t)]$$

Hence,

$$x_h = \int_0^{T_s} s_1(t) \cdot \phi_1(t)\, dt = \int_0^{T_s} \sqrt{\frac{E_s}{T_s}} [\cos(2\pi f_c t) - \sin(2\pi f_c t)]$$

$$\cdot \sqrt{\frac{2}{T_s}} \cos(2\pi f_c t)\, dt$$

$$= \frac{\sqrt{2E_s}}{T_s} \int_0^{T_s} \left[\cos^2(2\pi f_c t) - \cos(2\pi f_c t) \sin(2\pi f_c t)\right] dt$$

$$= \frac{\sqrt{2E_s}}{T_s} \left[\int_0^{T_s} \cos^2(2\pi f_c t) dt - \int_0^{T_s} \cos(2\pi f_c t) \sin(2\pi f_c t) dt\right]$$

We know

$$\int_0^{T_s} \cos^2(2\pi f_c t) dt = \frac{T_s}{2}, \quad \int_0^{T_s} \cos(2\pi f_c t) \sin(2\pi f_c t) dt = 0$$

$$\Rightarrow x = \frac{\sqrt{2E_s}}{T_s} \cdot \frac{T_s}{2} = \sqrt{\frac{E_s}{2}}$$

$$x_v = \int_0^{T_s} s_1(t) \cdot \phi_2(t)\, dt = \int_0^{T_s} \sqrt{\frac{E_s}{T_s}} [\cos(2\pi f_c t) - \sin(2\pi f_c t)]$$

$$\cdot \sqrt{\frac{2}{T_s}} \sin(2\pi f_c t)\, dt$$

$$= \frac{\sqrt{2E_s}}{T_s} \left[\int_0^{T_s} \cos(2\pi f_c t) \sin(2\pi f_c t) dt - \int_0^{T_s} \sin^2(2\pi f_c t) dt\right]$$

$$= \frac{\sqrt{2E_s}}{T_s} \left[0 - \frac{T_s}{2}\right] = -\sqrt{\frac{E_s}{2}}$$

therefore the coordinates of signal point are

$$\boxed{\left(\sqrt{\frac{E_s}{2}}, -\sqrt{\frac{E_s}{2}}\right)}$$

An alternative approach is to express the signal $s_1(t)$ in terms of the basis functions $\phi_1(t)$ and $\phi_2(t)$ and make use of the orthogonality between these functions.

(ii)

$$r = \sqrt{x^2 + y^2} = \sqrt{\left(\sqrt{\frac{E_s}{2}}\right)^2 + \left(-\sqrt{\frac{E_s}{2}}\right)^2} = \sqrt{\frac{E_s}{2} + \frac{E_s}{2}}$$

$$\boxed{r = \sqrt{E_s}}$$

[7]

(i) In M-ary Phase-Shift Keying (PSK), $\log_2 M$ code digits are represented by a signal $s_i(t)$ of time duration T_s given by

$$s_i(t) = A\cos\left(2\pi f_c t + \frac{2\pi i}{M}\right)$$

where f_c is the carrier frequency, and A is the amplitude of the signal. Each symbol represents $\log_2 M$ bits. Let the input binary rate be r_b (bits/sec).

Uncoded

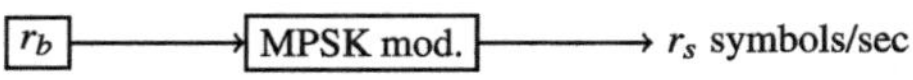

$$r_s = \frac{r_b}{\log_2 M} \quad \Rightarrow \quad T_s = \frac{1}{r_s} = \frac{\log_2 M}{r_b} = T_b \log_2 M$$

Coded

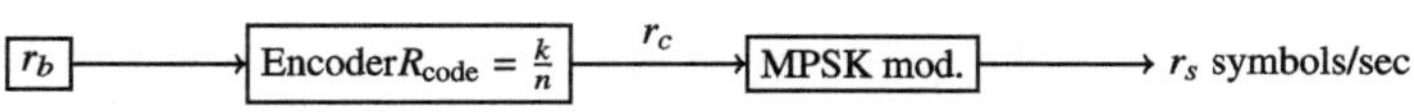

Let r_c be the code digits/sec output from the encoder. Then

$$R_{\text{code}} = \frac{r_b}{r_c} \quad \Rightarrow \quad r_c = \frac{r_b}{R_{\text{code}}}$$

$$r_s = \frac{r_c}{\log_2 M} = \frac{r_b}{R_{\text{code}} \log_2 M} \quad \Rightarrow \quad T_s = \frac{1}{r_s} = \frac{R_{\text{code}} \log_2 M}{r_b} = T_b R_{\text{code}} \log_2 M$$

$$\boxed{T_s = T_b R_{\text{code}} \log_2 M}$$

Given that the amplitude of an MPSK signal is the same as that of a BPSK signal because the information is represented within the phase only,

$$\text{MPSK Amplitude} = \text{BPSK Amplitude}$$

Thus,

$$\sqrt{\frac{2E_s}{T_s}} = \sqrt{\frac{2E_b}{T_b}}$$

Squaring both sides

$$\frac{E_s}{T_s} = \frac{E_b}{T_b}$$

$$E_s = E_b \cdot \frac{T_s}{T_b}$$

Using the result from part (i)

$$T_s = T_b R_{\text{code}} \log_2 M$$

$$E_s = E_b \cdot \frac{T_b R_{\text{code}} \log_2 M}{T_b} = E_b R_{\text{code}} \log_2 M$$

$$\boxed{E_s = E_b R_{\text{code}} \log_2 M}$$

(ii) For $M = 4$ (QPSK), the signal points lie on a circle of radius $\sqrt{E_s}$. The coordinates of each point are

$$\left(\pm\sqrt{\frac{E_s}{2}}, \pm\sqrt{\frac{E_s}{2}}\right)$$

The diagrams show the signal points for $E_b = 1$ Joule.

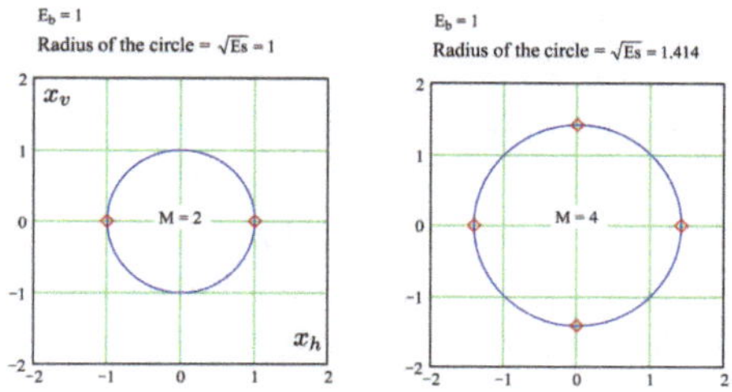

(iii) Probability of correctly receiving symbol 11

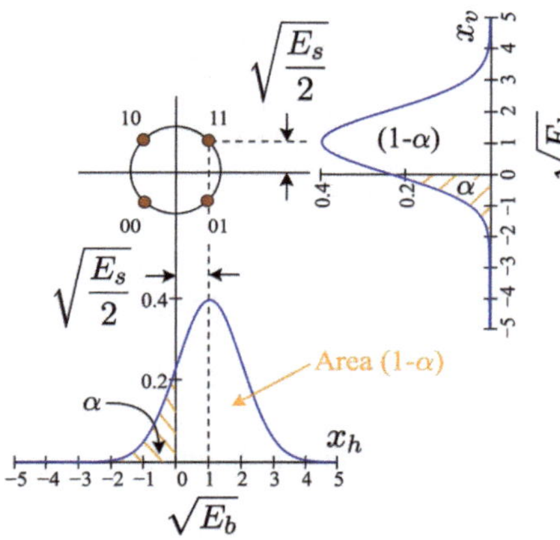

$$P_c = P(x_h \geq 0 \text{ AND } x_v \geq 0) = (1-\alpha)(1-\alpha)$$

Hence the probability of a symbol error is given by

$$P_s = 1 - P_c = 1 - (1-\alpha)^2 = 1 - (1 - 2\alpha + \alpha^2) = 2\alpha - \alpha^2$$

Using Gray coding, where only one bit changes between adjacent constellation points, the probability of a bit error is

$$P_e = \frac{P_s}{2} = \frac{2\alpha - \alpha^2}{2}$$

$$\boxed{P_e = \frac{2\alpha - \alpha^2}{2}}, \quad \text{where } \alpha = Q\left(\sqrt{\frac{2E_b}{N_0}}\right)$$

[8]

BPSK

BPSK transmits one binary digit per symbol, and the constellation points are

$$\pm\sqrt{E_b}$$

The minimum Euclidean distance between symbols is

$$d_{\text{min,BPSK}} = 2\sqrt{E_b}$$

and the bit error probability is

$$P_e^{\text{BPSK}} = Q\left(\sqrt{\frac{2E_b}{N_0}}\right)$$

8-PSK

In 8-PSK, the signal constellation consists of eight equally spaced points on a circle of radius $\sqrt{E_s}$. Each symbol represents three binary digits, thus for uncoded 8-PSK

$$E_s = 3E_b$$

The angular separation between adjacent points is

$$\theta = \frac{2\pi}{8} = \frac{\pi}{4}$$

Using the law of cosines, the minimum Euclidean distance between adjacent points is

$$d_{\text{min,8PSK}} = \sqrt{6E_b(1-\cos(\theta))} = \sqrt{6E_b\left(1-\frac{1}{\sqrt{2}}\right)}$$

This is clearly **less** than $d_{\text{min,BPSK}}$ for the same bit energy, since

$$6\left(1-\frac{1}{\sqrt{2}}\right) < 2$$

The approximate symbol error probability for MPSK for $M \geq 4$ is given by

$$P_s^{\text{MPSK}} \approx 2Q\left(\sqrt{\frac{2E_s}{N_0}}\cdot\sin\left(\frac{\pi}{M}\right)\right)$$

For $M = 8$, this becomes:

$$P_s^{\text{8PSK}} \approx 2Q\left(\sqrt{\frac{2 \cdot 3E_b}{N_0}} \cdot \sin\left(\frac{\pi}{8}\right)\right)$$

The factor of 2 and $\sin\left(\frac{\pi}{8}\right) \approx 0.383$ makes $P_e^{\text{8PSK}} > P_e^{\text{BPSK}}$ for a given SNR. The probability of error in 8-PSK is worse than that in BPSK because the constellation points in 8-PSK are more closely spaced, reducing the minimum distance and increasing the likelihood of symbol confusion under noise, as also shown in the diagram below. Note that all signal points are sent many times under various SNRs to show the level of overlap between adjacent sectors.

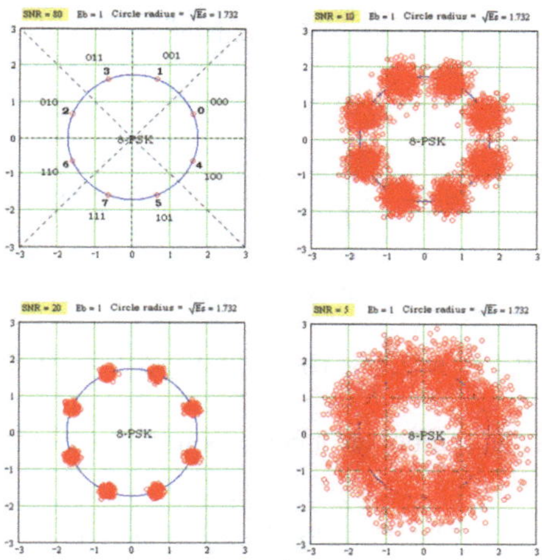

[9]

We are given

- $M = 16 \Rightarrow \log_2 M = 4$
- $r_b = 1000$ binary digits/sec
- $A = 1.5\text{ mV} = 1.5 \times 10^{-3}\text{ V}$
- $N_0 = 1 \times 10^{-11}$ W/Hz

For uncoded 16-PSK

$$P_s = 2Q\left(\sqrt{\frac{2E_s}{N_0}} \cdot \sin\left(\frac{\pi}{M}\right)\right)$$

The amplitude of the carrier

$$A = \sqrt{\frac{2E_s}{T_s}} \quad \Rightarrow \quad T_s = \frac{2E_s}{A^2}$$

But,

$$T_s = T_b \log_2 M = \frac{4}{r_b} \quad \Rightarrow \quad E_s = \frac{A^2 T_s}{2} = \frac{A^2}{2} \cdot \frac{4}{r_b} = \frac{2A^2}{r_b}$$

$$E_s = \frac{2 \cdot (1.5 \times 10^{-3})^2}{1000} = 4.5 \times 10^{-9} \text{ J}$$

Using the approximation

$$Q(u) \approx \frac{1}{2\sqrt{\pi}} e^{-u^2/2}$$

$$P_s \approx 2.1 \times 10^{-8}$$

The symbol rate

$$r_s = \frac{1}{T_s} = \frac{r_b}{\log_2 M} = \frac{1000}{4} = 250 \text{ symbols/sec}$$

Thus, the number of symbol errors per second

$$P_s \cdot r_s = 2.1 \times 10^{-8} \cdot 250 = 5.1 \times 10^{-6}$$

With Gray coding, one symbol error corresponds to one binary digit error. Hence:

$$\text{Bit errors/sec} \approx \boxed{5.1 \times 10^{-6}}$$

or equivalently, one error every

$$\frac{1}{5.1 \times 10^{-6}} \approx 1.96 \times 10^5 \text{ seconds} \approx 2.25 \text{ days}$$

However, if $r_b = 2855$ binary digits per second, then we can expect a binary digit error every second.

[10] To determine the probability of error when using a repetition code of rate $\frac{1}{N}$ over a Binary Symmetric Channel (BSC) with crossover probability α, consider first $N = 3$, where each bit is transmitted three times. The decoder uses majority voting to decide the original bit. That means an error occurs if at least two out of three bits are received in error. To determine the probability of error, consider the following cases:

- **Exactly 2 errors**
  Assume 000 transmitted. All possible error patterns are 011, 101, 110.
  Probability = $\binom{3}{2}\alpha^2(1-\alpha) = 3\alpha^2(1-\alpha)$

- **Exactly 3 errors**
  Assume 000 transmitted. The only possible error pattern is 111.
  Probability = $\binom{3}{3}\alpha^3 = \alpha^3$

  Thus, the total probability of error is

$$P_{\text{rep}}(3, \alpha) = 3\alpha^2(1 - \alpha) + \alpha^3$$

In general, let N be any odd number. An error occurs if **more than half** the bits are flipped during transmission. That is, the decoder makes a wrong decision if the number of bit errors i satisfies

$$i \geq \left\lfloor \frac{N}{2} \right\rfloor + 1 = \left(\frac{N-1}{2} + 1 \right)$$

The probability of exactly i errors is

$$P(i \text{ errors}) = \binom{N}{i} \alpha^i (1 - \alpha)^{N-i}$$

Hence, the total probability of error is

$$P_{\text{rep}}(N, \alpha) = \sum_{i=\left(\frac{N-1}{2}+1\right)}^{N} \binom{N}{i} \alpha^i (1 - \alpha)^{N-i}$$

[11] Let x_1 represent the binary digit 0 and x_2 represent the binary digit 1. Let α be the BSC crossover probability. Let p be the probability with which x_1 is produced, i.e., $P(x_1) = p, \quad P(x_2) = 1 - p$. The binary symmetric channel has the following conditional probabilities:

y_1	$P(y_1 \mid x_1) = 1 - \alpha$	$P(y_1 \mid x_2) = \alpha$
y_2	$P(y_2 \mid x_1) = \alpha$	$P(y_2 \mid x_2) = 1 - \alpha$

The entropy of the source is given by

$$H(X) = P(x_1) \log_2 \frac{1}{P(x_1)} + P(x_2) \log_2 \frac{1}{P(x_2)}$$

$$H(X) = p \log_2 \frac{1}{p} + (1 - p) \log_2 \frac{1}{(1 - p)} = \Omega(p) \quad \text{bits/symbol}$$

Hence, the information rate R from the source into the BSC is

$$R = r_b \cdot \Omega(p) \quad \text{bits/sec}$$

To determine the output entropy $H(Y)$, we require

$$P(y_1) = P(y_1|x_1)P(x_1) + P(y_1|x_2)P(x_2) = (1-\alpha)p + \alpha(1-p) = p + \alpha - 2\alpha p$$

$$P(y_2) = 1 - P(y_1)$$

Then

$$\begin{aligned} H(Y) &= P(y_1)\log_2 \frac{1}{P(y_1)} + (1 - P(y_1))\log_2 \frac{1}{1 - P(y_1)} \\ &= \Omega(p + \alpha - 2\alpha p) \quad \text{bits/symbol} \end{aligned}$$

The conditional entropy

$$H(Y|X) = \sum_{i=1}^{2}\sum_{j=1}^{2} P(y_j|x_i)P(x_i)\log_2 \frac{1}{P(y_j|x_i)}$$

$$\begin{aligned} H(Y|X) = (1-\alpha)P(x_1)\log_2 \frac{1}{1-\alpha} + \alpha P(x_1)\log_2 \frac{1}{\alpha} + \alpha P(x_2)\log_2 \frac{1}{\alpha} \\ + (1-\alpha)P(x_2)\log_2 \frac{1}{1-\alpha} \end{aligned}$$

$$= P(x_1)\cdot\Omega(\alpha) + P(x_2)\cdot\Omega(\alpha) = \Omega(\alpha)[P(x_1) + P(x_2)] = \Omega(\alpha)$$

Hence, the average mutual information is given by

$$I(X;Y) = H(Y) - H(Y|X) = \Omega(p + \alpha - 2\alpha p) - \Omega(\alpha) \quad \text{bits/symbol}$$

where we maximize the mutual information over all p to determine the channel capacity

$$C_s = \max_p\,[\Omega(p + \alpha - 2\alpha p) - \Omega(\alpha)]$$

The binary entropy function $\Omega(z)$ is maximized at $z = \frac{1}{2}$, so we set

$$p + \alpha - 2\alpha p = \frac{1}{2}$$

which is satisfied for $p = 1/2$. Thus,

$$C_s = \Omega\left(\frac{1}{2}\right) - \Omega(\alpha) = 1 - \Omega(\alpha) \quad \text{bits/symbol}$$

The channel capacity in bits/sec is given by

$$C = r_b(1 - \Omega(\alpha)) \quad \text{bits/sec}$$

- If $\alpha = 0$ the channel is noiseless and $C = r_b = R$ as expected
- If $\alpha = \frac{1}{2}$, the channel is completely noisy and $C = 0$

[12]

(i) The probabilities given are

$$P(y_1|x_1) = P(y_3|x_2) = p, \quad P(y_2|x_1) = P(y_2|x_2) = 1 - p$$

Assuming equiprobable inputs

$$P(x_1) = P(x_2) = \frac{1}{2}$$

We find the probabilities for outputs

$$P(y_1) = P(x_1)P(y_1|x_1) + P(x_2)P(y_1|x_2) = \frac{1}{2}p + \frac{1}{2}(0) = \frac{p}{2}$$

$$P(y_2) = P(x_1)P(y_2|x_1) + P(x_2)P(y_2|x_2) = \frac{1}{2}(1-p) + \frac{1}{2}(1-p) = 1 - p$$

$$P(y_3) = P(x_1)P(y_3|x_1) + P(x_2)P(y_3|x_2) = \frac{1}{2}(0) + \frac{1}{2}p = \frac{p}{2}$$

Entropy

$$H(Y) = \sum_j P(y_j)\log_2 \frac{1}{P(y_j)}$$

$$= P(y_1)\log_2 \frac{1}{P(y_1)} + P(y_2)\log_2 \frac{1}{P(y_2)} + P(y_3)\log_2 \frac{1}{P(y_3)}$$

$$= \frac{p}{2}\log_2 \frac{2}{p} + (1-p)\log_2 \frac{1}{1-p} + \frac{p}{2}\log_2 \frac{2}{p}$$

$$= p\log_2\frac{2}{p} + (1-p)\log_2\frac{1}{1-p}$$

$$= p\log_2 2 + p\log_2\frac{1}{p} + (1-p)\log_2\frac{1}{1-p}$$

$$= p + \Omega(p)$$

(ii) Given

$$H(Y|X) = \Omega(p)$$

The mutual information is

$$I(X;Y) = H(Y) - H(Y|X) = p + \Omega(p) - \Omega(p) = p$$

Hence, the channel capacity

$$\boxed{C_s = I(X;Y) = p \text{ bits/symbol}}$$

[13]

(i) Given

$$P(x_2) = P(x_3)$$

let

$$P(x_1) = p$$

Thus,

$$p + 2P(x_2) = 1 \Rightarrow P(x_2) = P(x_3) = \frac{1-p}{2}$$

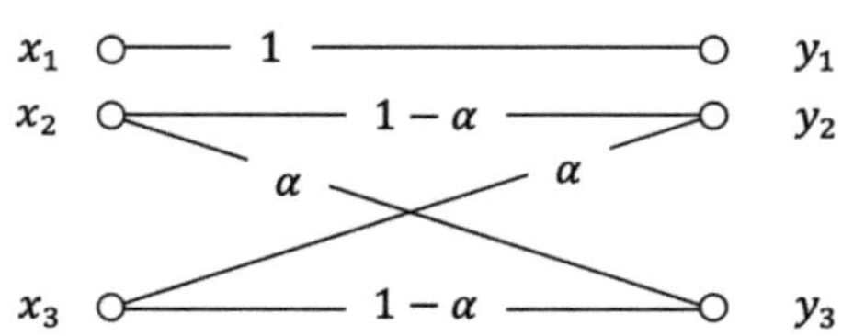

Now

$$P(y_1) = P(x_1) = p$$

$$P(y_2) = P(x_2)(1-\alpha) + P(x_3)(\alpha) = \frac{1-p}{2}(1-\alpha) + \frac{1-p}{2}(\alpha) = \frac{1-p}{2}$$

$$P(y_3) = P(x_2)(\alpha) + P(x_3)(1-\alpha) = \frac{1-p}{2}(\alpha) + \frac{1-p}{2}(1-\alpha) = \frac{1-p}{2}$$

Thus,

$$H(Y) = \sum_j P(y_j)\log_2\frac{1}{P(y_j)}$$

$$= p\log_2\frac{1}{p} + \frac{1-p}{2}\log_2\frac{2}{1-p} + \frac{1-p}{2}\log_2\frac{2}{1-p}$$

$$= p\log_2\frac{1}{p} + (1-p)\log_2\frac{2}{1-p}$$

$$= p\log_2\frac{1}{p} + (1-p)\left[1 + \log_2\frac{1}{1-p}\right]$$

$$= \boxed{\Omega(p) + 1 - p}$$

(ii)

$$H(Y|X) = \sum_i\sum_j P(y_j|x_i)\,P(x_i)\,\log_2\left(\frac{1}{P(y_j|x_i)}\right)$$

Substitute the known values. Since $P(y_1|x_1) = 1$ contributes zero entropy and the pairs (x_2, x_3) each pass through a BSC with crossover probability α,

$$H(Y|X) = (1-p)\left[(1-\alpha)\log_2\left(\frac{1}{1-\alpha}\right) + \alpha\log_2\left(\frac{1}{\alpha}\right)\right] = (1-p)\Omega(\alpha).$$

Hence, the average mutual information is

$$\begin{aligned} I(X;Y) &= H(Y) - H(Y|X) \\ &= \Omega(p) + 1 - p - (1-p)\Omega(\alpha) \\ &= \boxed{\Omega(p) + (1-p)\left[1 - \Omega(\alpha)\right]}. \end{aligned}$$

(iii) The average mutual information equals the source entropy when $H(X|Y) = 0$, i.e., when X is a deterministic function of Y (or when the source is degenerate). For this DMC, this occurs when the channel is deterministic on the $\{x_2, x_3\}$ branch, i.e.,

$$\alpha \in \{0, 1\},$$

or when $p = 1$ (so that $H(X) = 0$). For $\alpha \in \{0, 1\}$ we have $H(Y|X) = (1-p)\Omega(\alpha) = 0$ and Y uniquely determines X, hence

$$I(X;Y) = H(X).$$

[14] Diagram of the PDF, quantization levels, and forward transition probabilities.

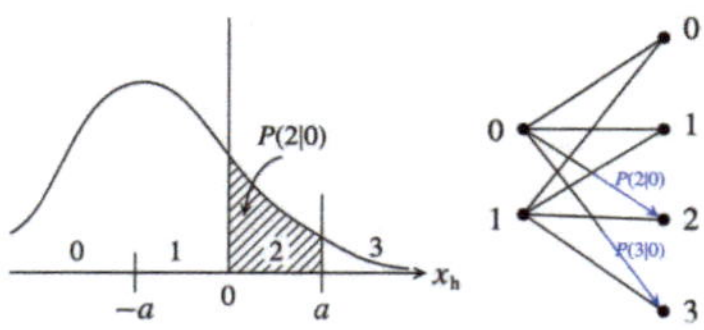

(i)

$$f(x_h|0) = \frac{1}{\sigma\sqrt{2\pi}} \exp\left(-\frac{(x_h + \sqrt{E_b})^2}{2\sigma^2}\right)$$

$$P(2|0) = \int_{x_h=0}^{x_h=a} \frac{1}{\sigma\sqrt{2\pi}} \exp\left(-\frac{(x_h + \sqrt{E_b})^2}{2\sigma^2}\right) dx_h$$

$$\text{Let } z = \frac{x_h + \sqrt{E_b}}{\sigma} \Rightarrow \frac{dz}{dx_h} = \frac{1}{\sigma} \Rightarrow \sigma dz = dx_h$$

$$\text{For } x_h = 0, \quad z = \frac{\sqrt{E_b}}{\sigma}$$

$$\text{For } x_h = a, \quad z = \frac{a + \sqrt{E_b}}{\sigma}$$

$$\Rightarrow P(2|0) = \int_{\frac{\sqrt{E_b}}{\sigma}}^{\frac{a+\sqrt{E_b}}{\sigma}} \frac{1}{\sqrt{2\pi}} e^{-z^2/2} dz = Q\left(\frac{\sqrt{E_b}}{\sigma}\right) - Q\left(\frac{a+\sqrt{E_b}}{\sigma}\right)$$

$$P(3|0) = \int_{x_h=a}^{\infty} \frac{1}{\sigma\sqrt{2\pi}} e^{-\frac{(x_h+\sqrt{E_b})^2}{2\sigma^2}} dx_h$$

Change of variable: $z = \dfrac{x_h + \sqrt{E_b}}{\sigma} \Rightarrow x_h = \sigma z - \sqrt{E_b}$

For $x_h = a, \quad z = \dfrac{a + \sqrt{E_b}}{\sigma}$

$$\Rightarrow P(3|0) = \int_{\frac{a+\sqrt{E_b}}{\sigma}}^{\infty} \frac{1}{\sqrt{2\pi}} e^{-z^2/2} dz = Q\left(\frac{a+\sqrt{E_b}}{\sigma}\right)$$

$\therefore P(3|0) + P(2|0) = \alpha$ (expected)

Let's check

$$Q\left(\frac{a+\sqrt{E_b}}{\sigma}\right) + Q\left(\frac{\sqrt{E_b}}{\sigma}\right) - Q\left(\frac{a+\sqrt{E_b}}{\sigma}\right) = Q\left(\frac{\sqrt{E_b}}{\sigma}\right)$$

Which for $\sigma^2 = \dfrac{N_0}{2}$ gives:

$\Rightarrow \alpha = Q\left(\sqrt{\dfrac{2E_b}{N_0}}\right)$ As expected for uncoded BPSK.

(ii)

$$\left(\frac{E_b}{N_0}\right)_{\text{dB}} = 10\log_{10}\left(\frac{E_b}{N_0}\right) = 0 \text{ dB}$$

$$\Rightarrow \frac{E_b}{N_0} = 10^{0/10} = 10^0 = 1$$

$$\alpha = Q(\sqrt{2}) \approx 0.1$$

For this simple BSC, $P_e = \alpha = 0.1$

[15] Consider

$$P(4|0) = \int_0^a f(x_h|0)\, dx_h$$

$$= \int_0^a \frac{1}{\sigma\sqrt{2\pi}} \exp\left[-\frac{(x_h + \sqrt{E_b})^2}{2\sigma^2}\right] dx_h$$

To integrate by substitution, let

$$z = \frac{x_h + \sqrt{E_b}}{\sigma} \quad \text{Thus} \quad \frac{dz}{dx_h} = \frac{1}{\sigma}$$

For $x_h = 0$, $z = \frac{\sqrt{E_b}}{\sigma}$ and for $x_h = a$, $z = \frac{a+\sqrt{E_b}}{\sigma}$
Hence,

$$P(4|0) = \int_{\frac{\sqrt{E_b}}{\sigma}}^{\frac{a+\sqrt{E_b}}{\sigma}} \frac{1}{\sqrt{2\pi}} \exp\left[-\frac{z^2}{2}\right] dz$$

Notice how the σ canceled. Since

$$Q(u) = \int_u^{\infty} \frac{1}{\sqrt{2\pi}} \exp\left(-\frac{z^2}{2}\right) dz$$

We find

$$P(4|0) = Q\left(\frac{\sqrt{E_b}}{\sigma}\right) - Q\left(\frac{a+\sqrt{E_b}}{\sigma}\right)$$

Following a similar procedure, the forward transition probabilities for a binary digit 0 transmitted are given by

$$P(4|0) = \int_0^{a} f(x_h|0)\, dx_h = Q\left(\frac{\sqrt{E_b}}{\sigma}\right) - Q\left(\frac{a+\sqrt{E_b}}{\sigma}\right)$$

$$P(5|0) = \int_a^{2a} f(x_h|0)\, dx_h = Q\left(\frac{a+\sqrt{E_b}}{\sigma}\right) - Q\left(\frac{2a+\sqrt{E_b}}{\sigma}\right)$$

$$P(6|0) = \int_{2a}^{3a} f(x_h|0)\, dx_h = Q\left(\frac{2a+\sqrt{E_b}}{\sigma}\right) - Q\left(\frac{3a+\sqrt{E_b}}{\sigma}\right)$$

$$P(7|0) = \int_{3a}^{\infty} f(x_h|0)\, dx_h = Q\left(\frac{3a+\sqrt{E_b}}{\sigma}\right)$$

Since each PDF integrates to 1,

$$\sum_{j=0}^{7} P(j|0) = \int_{-\infty}^{\infty} f(x_h|0) dx_h = 1$$

and for example,

$$P(7|0) + P(6|0) + P(5|0) + P(4|0) = \int_0^\infty f(x_h|0)dx_h$$

$$= Q\left(\frac{\sqrt{E_b}}{\sigma}\right) = \boxed{\alpha}$$

Follow a similar procedure yourself to show that

$$\sum_{j=0}^{7} P(j|1) = \int_{-\infty}^{\infty} f(x_h|1)dx_h = 1$$

Appendix B
Solutions to Chapter 2

[1] The node functions are

$$g(a, b, i) = b + (1 - 2i)a \quad \Longrightarrow \quad \begin{cases} g(a, b, 0) = b + a \\ g(a, b, 1) = b - a \end{cases}$$

$$f(a, b) = \text{sign}(a)\ \text{sign}(b)\ \min\{|a|, |b|\}.$$

The hard-decision outputs of the BSC are denoted by $r_1, r_2, r_3, r_4 \in \{\pm 1\}$, where '+' denotes $+1$ (no flip) and '−' denotes -1 (a flip). Thus,

$$L_{23} = r_1 + r_3,$$

$$L_{24} = r_2 + r_4,$$

$$L_{33} = f(L_{23}, L_{24}) = \text{sign}(L_{23})\ \text{sign}(L_{24}) \min\big(|L_{23}|, |L_{24}|\big).$$

The SC decision on u_3 is

$$\hat{u}_3 = \begin{cases} 0 & L_{33} \geq 0, \\ 1 & L_{33} < 0. \end{cases}$$

J. Sodha, *Fundamentals of Polar, LDPC and Turbo Codes*,
https://doi.org/10.1007/978-3-031-90924-5

r_1	r_2	r_3	r_4	L_{23}	L_{24}	L_{33}	$\hat{u}_3$
+	+	+	+	2	2	+2	0
+	+	+	−	2	0	0	0
+	+	−	+	0	2	0	0
+	+	−	−	0	0	0	0
+	−	+	+	2	0	0	0
+	−	+	−	2	−2	−2	1
+	−	−	+	0	0	0	0
+	−	−	−	0	−2	0	0
−	+	+	+	0	2	0	0
−	+	+	−	0	0	0	0
−	+	−	+	−2	2	−2	1
−	+	−	−	−2	0	0	0
−	−	+	+	0	0	0	0
−	−	+	−	0	−2	0	0
−	−	−	+	−2	0	0	0
−	−	−	−	−2	−2	+2	0

A wrong decision ($\hat{u}_3 = 1$) occurs only when $(r_1 + r_3)$ and $(r_2 + r_4)$ are nonzero and have opposite signs, i.e.,

$$(+, -, +, -) \text{ and } (-, +, -, +).$$

Each pattern has probability $\alpha^2(1 - \alpha)^2$, hence

$$\boxed{P_e(u_3) = 2\alpha^2\big(1 - \alpha\big)^2} \qquad (0 \le \alpha \le \tfrac{1}{2}).$$

Assuming the use of BPSK over an AWGN channel with hard decisions,

$$\alpha = Q\Big(\sqrt{2\, R_{\text{code}}\, E_b/N_0}\Big) = Q\Big(\sqrt{2R_{\text{code}}\, 10^{\text{SNR}/10}}\Big).$$

To verify the theoretical formula by simulation, set $E_b = \frac{1}{R_{\text{code}}}$ so that the transmitted signal points lie at ± 1. Then

$$\sigma = \sqrt{\frac{E_b}{2 \cdot 10^{\text{SNR}/10}}}$$

may be used in the Gaussian noise generator.

[2]

Refer to the polar tree shown for the solution.

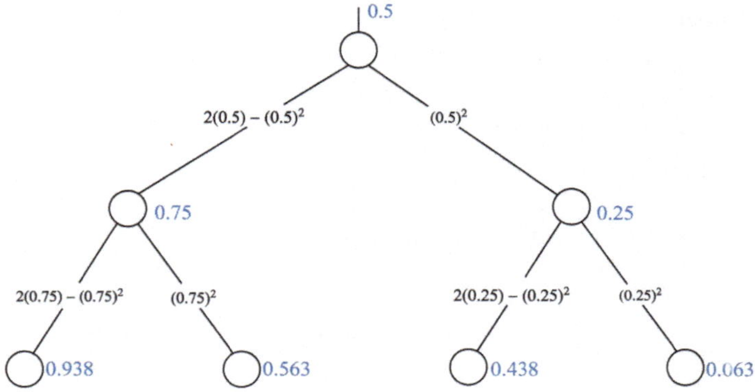

[3] All $2^4 = 16$ codewords of the (8, 4) polar code with information bits placed in (u_4, u_6, u_7, u_8) and the remaining u_i frozen to 0 are shown in the table below.

	Message				Codeword								
Row	u_4	u_6	u_7	u_8	c_1	c_2	c_3	c_4	c_5	c_6	c_7	c_8	Weight w
1	0	0	0	0	0	0	0	0	0	0	0	0	0
2	0	0	0	1	1	1	1	1	1	1	1	1	8
3	0	0	1	0	1	0	1	0	1	0	1	0	4
4	0	0	1	1	0	1	0	1	0	1	0	1	4
5	0	1	0	0	1	1	0	0	1	1	0	0	4
6	0	1	0	1	0	0	1	1	0	0	1	1	4
7	0	1	1	0	0	1	1	0	0	1	1	0	4
8	0	1	1	1	1	0	0	1	1	0	0	1	4
9	1	0	0	0	1	1	1	1	0	0	0	0	4
10	1	0	0	1	0	0	0	0	1	1	1	1	4
11	1	0	1	0	0	1	0	1	1	0	1	0	4
12	1	0	1	1	1	0	1	0	0	1	0	1	4
13	1	1	0	0	0	0	1	1	1	1	0	0	4
14	1	1	0	1	1	1	0	0	0	0	1	1	4
15	1	1	1	0	1	0	0	1	0	1	1	0	4
16	1	1	1	1	0	1	1	0	1	0	0	1	4

Note that the minimum distance $d_{\min} = 4$. The code can therefore correct any single bit error or detect up to three errors.

[4]

Code Parameters

- Block length: $N = 8$
- Message length: $K = 4$
- Code rate: $R_{\text{code}} = \frac{K}{N} = 0.5$
- SNR: 3 dB, resulting in $\sigma = 0.708$
- Frozen positions: $\{1, 2, 3, 5\}$
- Message inserted at positions u_4, u_6, u_7, u_8
- Received LLR vector:

$$r = [1.2, -0.8, 0.6, -1.4, 1.5, -1.0, 0.4, -1.2]$$

SC Decoding Algorithm

Node functions are as follows.

$$f(a, b) = \text{sign}(a)\,\text{sign}(b)\min(|a|, |b|)$$
$$g(a, b, u) = b + (1 - 2u) \cdot a$$

Step 0: Initialization with r

$$L = \begin{bmatrix} 1.2 & -0.8 & 0.6 & -1.4 & 1.5 & -1.0 & 0.4 & -1.2 \\ 0 & 0 & 0 & 0 & 0 & 0 & 0 & 0 \\ 0 & 0 & 0 & 0 & 0 & 0 & 0 & 0 \\ 0 & 0 & 0 & 0 & 0 & 0 & 0 & 0 \end{bmatrix}$$

Step 1: Compute $L_{2,1}$ to $L_{2,4}$ using f

$$L_{2,1} = f(1.2, 1.5) = 1.2, L_{2,2} = f(-0.8, -1.0) = 0.8,$$
$$L_{2,3} = f(0.6, 0.4) = 0.4, L_{2,4} = f(-1.4, -1.2) = 1.2$$

$$L = \begin{bmatrix} 1.2 & -0.8 & 0.6 & -1.4 & 1.5 & -1.0 & 0.4 & -1.2 \\ 1.2 & 0.8 & 0.4 & 1.2 & 0 & 0 & 0 & 0 \\ 0 & 0 & 0 & 0 & 0 & 0 & 0 & 0 \\ 0 & 0 & 0 & 0 & 0 & 0 & 0 & 0 \end{bmatrix}$$

Step 2: $L_{3,1} = f(1.2, 0.4) = 0.4$, $L_{3,2} = f(0.8, 1.2) = 0.8$

$$L = \begin{bmatrix} 1.2 & -0.8 & 0.6 & -1.4 & 1.5 & -1.0 & 0.4 & -1.2 \\ 1.2 & 0.8 & 0.4 & 1.2 & 0 & 0 & 0 & 0 \\ 0.4 & 0.8 & 0 & 0 & 0 & 0 & 0 & 0 \\ 0 & 0 & 0 & 0 & 0 & 0 & 0 & 0 \end{bmatrix}$$

Step 3: $L_{4,1} = f(0.4, 0.8) = 0.4$, **frozen decision** $u_1 = 0$

$$L = \begin{bmatrix} 1.2 & -0.8 & 0.6 & -1.4 & 1.5 & -1.0 & 0.4 & -1.2 \\ 1.2 & 0.8 & 0.4 & 1.2 & 0 & 0 & 0 & 0 \\ 0.4 & 0.8 & 0 & 0 & 0 & 0 & 0 & 0 \\ 0.4 & 0 & 0 & 0 & 0 & 0 & 0 & 0 \end{bmatrix}$$

Step 4: $L_{4,2} = g(0.4, 0.8, 0) = 1.2$, **frozen decision** $u_2 = 0$

$$L = \begin{bmatrix} 1.2 & -0.8 & 0.6 & -1.4 & 1.5 & -1.0 & 0.4 & -1.2 \\ 1.2 & 0.8 & 0.4 & 1.2 & 0 & 0 & 0 & 0 \\ 0.4 & 0.8 & 0 & 0 & 0 & 0 & 0 & 0 \\ 0.4 & 1.2 & 0 & 0 & 0 & 0 & 0 & 0 \end{bmatrix}$$

Step 5: $L_{3,3} = g(1.2, 0.4, 0) = 1.6$, $L_{3,4} = g(0.8, 1.2, 0) = 2.0$

$$L = \begin{bmatrix} 1.2 & -0.8 & 0.6 & -1.4 & 1.5 & -1.0 & 0.4 & -1.2 \\ 1.2 & 0.8 & 0.4 & 1.2 & 0 & 0 & 0 & 0 \\ 0.4 & 0.8 & 1.6 & 2.0 & 0 & 0 & 0 & 0 \\ 0.4 & 1.2 & 0 & 0 & 0 & 0 & 0 & 0 \end{bmatrix}$$

Step 6: $L_{4,3} = f(1.6, 2.0) = 1.6$, **frozen decision** $u_3 = 0$

$$L = \begin{bmatrix} 1.2 & -0.8 & 0.6 & -1.4 & 1.5 & -1.0 & 0.4 & -1.2 \\ 1.2 & 0.8 & 0.4 & 1.2 & 0 & 0 & 0 & 0 \\ 0.4 & 0.8 & 1.6 & 2.0 & 0 & 0 & 0 & 0 \\ 0.4 & 1.2 & 1.6 & 0 & 0 & 0 & 0 & 0 \end{bmatrix}$$

Step 7: $L_{4,4} = g(1.6, 2.0, 0) = 3.6$, **decision** $u_4 = 0$

$$L = \begin{bmatrix} 1.2 & -0.8 & 0.6 & -1.4 & 1.5 & -1.0 & 0.4 & -1.2 \\ 1.2 & 0.8 & 0.4 & 1.2 & 0 & 0 & 0 & 0 \\ 0.4 & 0.8 & 1.6 & 2.0 & 0 & 0 & 0 & 0 \\ 0.4 & 1.2 & 1.6 & 3.6 & 0 & 0 & 0 & 0 \end{bmatrix}$$

Step 8: Compute $L_{2,5}$ **to** $L_{2,8}$

$$L_{2,5} = g(1.5, 1.2, 0) = 2.7, \qquad L_{2,6} = g(-0.8, -1.0, 0) = -1.8,$$

$$L_{2,7} = g(0.6, 0.4, 0) = 1.0, \qquad L_{2,8} = g(-1.4, -1.2, 0) = -2.6$$

$$L = \begin{bmatrix} 1.2 & -0.8 & 0.6 & -1.4 & 1.5 & -1.0 & 0.4 & -1.2 \\ 1.2 & 0.8 & 0.4 & 1.2 & 2.7 & -1.8 & 1.0 & -2.6 \\ 0.4 & 0.8 & 1.6 & 2.0 & 0 & 0 & 0 & 0 \\ 0.4 & 1.2 & 1.6 & 3.6 & 0 & 0 & 0 & 0 \end{bmatrix}$$

Step 9: $L_{3,5} = f(2.7, 1.0) = 1.0$, $L_{3,6} = f(-1.8, -2.6) = 1.8$

$$L = \begin{bmatrix} 1.2 & -0.8 & 0.6 & -1.4 & 1.5 & -1.0 & 0.4 & -1.2 \\ 1.2 & 0.8 & 0.4 & 1.2 & 2.7 & -1.8 & 1.0 & -2.6 \\ 0.4 & 0.8 & 1.6 & 2.0 & 1.0 & 1.8 & 0 & 0 \\ 0.4 & 1.2 & 1.6 & 3.6 & 0 & 0 & 0 & 0 \end{bmatrix}$$

Step 10: $L_{4,5} = f(1.0, 1.8) = 1.0$, **frozen decision** $u_5 = 0$ **Step 11:** $L_{4,6} = g(1.0, 1.8, 0) = 2.8$, **decision** $u_6 = 0$ **Step 12:** $L_{3,7} = g(2.7, 1.0, 0) = 3.7$, $L_{3,8} = g(-1.8, -2.6, 0) = -4.4$ **Step 13:** $L_{4,7} = f(3.7, -4.4) = -3.7$, **decision** $u_7 = 1$ **Step 14:** $L_{4,8} = g(3.7, -4.4, 1) = -8.1$, **decision** $u_8 = 1$

$$L = \begin{bmatrix} 1.2 & -0.8 & 0.6 & -1.4 & 1.5 & -1.0 & 0.4 & -1.2 \\ 1.2 & 0.8 & 0.4 & 1.2 & 2.7 & -1.8 & 1.0 & -2.6 \\ 0.4 & 0.8 & 1.6 & 2.0 & 1.0 & 1.8 & 3.7 & -4.4 \\ 0.4 & 1.2 & 1.6 & 3.6 & 1.0 & 2.8 & -3.7 & -8.1 \end{bmatrix}$$

Final Decoded Message

$$\boxed{\text{msg} = [0, 0, 1, 1]}$$

[5] Recall for a single node in the polar tree

$$(c_1, c_2) = (u_1 \oplus u_2,\ u_2)$$

Left synthetic channel
While decoding u_1, the decoder does *not* know u_2, hence

$$u_1 = c_1 \oplus c_2$$

is available *iff* both c_1 and c_2 are not erased (Table B.1).

Table B.1 Erasure patterns for left channel

c_1	c_2	Recover u_1?	Probability
e	e	No	p^2
e	$\neq e$	No	$p(1-p)$
$\neq e$	e	No	$(1-p)p$
$\neq e$	$\neq e$	Yes	$(1-p)^2$

Table B.2 Erasure patterns for right channel

c_1	c_2	Recover u_2?	Probability
e	e	No	p^2
e	$\neq e$	Yes	$p(1-p)$
$\neq e$	e	Yes	$(1-p)p$
$\neq e$	$\neq e$	Yes	$(1-p)^2$

Hence, the erasure probability of the left channel p_- is given by

$$p_- = p^2 + p(1-p) + (1-p)p = \boxed{2p - p^2}$$

Right synthetic channel

Now u_1 is known, so u_2 can be recovered unless *both* c_1 and c_2 are erased because (Table B.2)

$$u_2 = \begin{cases} c_2 & c_2 \neq e, \\ c_1 \oplus u_1, & c_2 = e,\ c_1 \neq e, \\ \text{erasure}, & c_2 = e,\ c_1 = e. \end{cases}$$

Hence, the erasure probability of the right channel p_+ is given by

$$p_+ = \boxed{p^2}$$

Notice that the right child is more reliable and the left child is less reliable than the original BEC(p), i.e.

$$p_+ = p^2 < p < 2p - p^2 = p_-$$

[6]

Let us observe r_n in an infinitesimally small interval $[r_n, r_n + \Delta r_n]$. The probability of r_n falling in this interval given $c_n = 0$ is approximately

$$P\big(r_n \in [r_n, r_n + \Delta r_n] \mid c_n = 0\big) \approx f(r_n \mid c_n = 0)\, \Delta r_n$$

where we make use of the fact that probability is the area under a probability density function (PDF). Applying Bayes' theorem,

$$P(A \mid B) = \frac{P(B \mid A)\, P(A)}{P(B)}$$

we write

$$P(c_n = 0 \mid r_n \in [r_n,\, r_n + \Delta r_n]) = \frac{P(r_n \in [r_n,\, r_n + \Delta r_n] \mid c_n = 0)\, P(c_n = 0)}{P(r_n \in [r_n,\, r_n + \Delta r_n])}$$

Using the approximation above, the numerator becomes

$$f(r_n \mid c_n = 0)\, \Delta r_n \cdot P(c_n = 0)$$

and the denominator is approximated by

$$P(r_n \in [r_n,\, r_n + \Delta r_n]) \approx f(r_n)\, \Delta r_n$$

Substituting, we find

$$\begin{aligned} P(c_n = 0 \mid r_n \in [r_n,\, r_n + \Delta r_n]) &= \frac{f(r_n \mid c_n = 0)\, \Delta r_n \cdot P(c_n = 0)}{f(r_n)\, \Delta r_n} \\ &= \frac{f(r_n \mid c_n = 0)\, P(c_n = 0)}{f(r_n)} \end{aligned}$$

Therefore, in the limit as $\Delta r_n \to 0$, this yields

$$P(c_n = 0 \mid r_n) = \frac{f(r_n \mid c_n = 0)\, P(c_n = 0)}{f(r_n)}$$

[7]

Given the conditional probability density functions

$$f(r_n \mid \sqrt{E_s}) = \frac{1}{\sigma\sqrt{2\pi}} \exp\left[-\frac{(r_n - \sqrt{E_s})^2}{2\sigma^2}\right]$$

$$f(r_n \mid -\sqrt{E_s}) = \frac{1}{\sigma\sqrt{2\pi}} \exp\left[-\frac{(r_n + \sqrt{E_s})^2}{2\sigma^2}\right]$$

The marginal distribution of r_n is

$$f(r_n) = f(r_n \mid \sqrt{E_s}) \cdot P(\sqrt{E_s}) + f(r_n \mid -\sqrt{E_s}) \cdot P(-\sqrt{E_s})$$

Assuming equal priors

$$P(\sqrt{E_s}) = P(-\sqrt{E_s}) = \frac{1}{2}$$

Using Bayes' theorem

$$P(c_n = 0 \mid r_n) = \frac{f(r_n \mid \sqrt{E_s}) \cdot \frac{1}{2}}{f(r_n)}$$

Let

$$a = \frac{(r_n - \sqrt{E_s})^2}{2\sigma^2}, \qquad b = \frac{(r_n + \sqrt{E_s})^2}{2\sigma^2}$$

Then

$$P(c_n = 0 \mid r_n) = \frac{e^{-a}}{e^{-a} + e^{-b}} = \frac{1}{1 + e^{a-b}}$$

Thus,

$$a - b = \frac{(r_n - \sqrt{E_s})^2 - (r_n + \sqrt{E_s})^2}{2\sigma^2}$$

$$= \frac{[r_n^2 - 2r_n\sqrt{E_s} + E_s] - [r_n^2 + 2r_n\sqrt{E_s} + E_s]}{2\sigma^2} = \frac{-4r_n\sqrt{E_s}}{2\sigma^2} = -\frac{2r_n\sqrt{E_s}}{\sigma^2}$$

Therefore,

$$P(c_n = 0 \mid r_n) = \frac{1}{1 + \exp\left(-\frac{2r_n\sqrt{E_s}}{\sigma^2}\right)}$$

Similarly,

$$P(c_n = 1 \mid r_n) = \frac{1}{1 + \exp\left(\frac{2r_n\sqrt{E_s}}{\sigma^2}\right)}$$

[8]
We are given the channel reliability

$$L_c = \frac{2\sqrt{E_s}}{\sigma^2}$$

and the conditional probabilities

$$P(c_n = 0 \mid r_n) = \frac{1}{1 + \exp(-L_c r_n)}, \qquad P(c_n = 1 \mid r_n) = \frac{1}{1 + \exp(L_c r_n)}$$

Hence, the log-likelihood ratio is

$$\lambda_n = \ln\left(\frac{P(c_n = 0 \mid r_n)}{P(c_n = 1 \mid r_n)}\right) = \ln\left(\frac{1 + \exp(L_c r_n)}{1 + \exp(-L_c r_n)}\right)$$

We now factor out $\exp(L_c r_n)$ from the numerator

$$= \ln\left(\frac{\exp(L_c r_n)\,(1 + \exp(-L_c r_n))}{1 + \exp(-L_c r_n)}\right) = \ln\left(\exp(L_c r_n)\right)$$

$$\boxed{\lambda_n = L_c r_n}$$

[9]

Let

$$x = \tanh(u) = \frac{e^{2u} - 1}{e^{2u} + 1}$$

Thus,

$$x(e^{2u} + 1) = e^{2u} - 1$$

$$xe^{2u} + x = e^{2u} - 1$$

$$1 + x = e^{2u}(1 - x)$$

Hence,

$$e^{2u} = \frac{1 + x}{1 - x} = \boxed{\frac{1 + \tanh(u)}{1 - \tanh(u)}}$$

[10]

The calculations follow the reliability sequence for the (8,4) code but now extend to $N = 16$ nodes. Refer to the tree shown below, where the Bhattacharyya parameters for $N = 16$ synthetic channels are listed at the leaf nodes. The Bhattacharyya parameter estimates the reliability of each bit channel, with lower values indicating more reliable channels (Table B.3).

Table B.3 Bhattacharyya Parameters for $N = 16$ Synthetic Channels

Bhattacharyya Parameter Z	Index
1	1
0.992	2
0.985	3
0.963	5
0.9	9
0.772	4
0.654	6
0.533	7
0.467	10
0.346	11
0.228	13
0.1	8
0.037	12
0.015	14
7.797×10^{-3}	15
1.526×10^{-5}	16

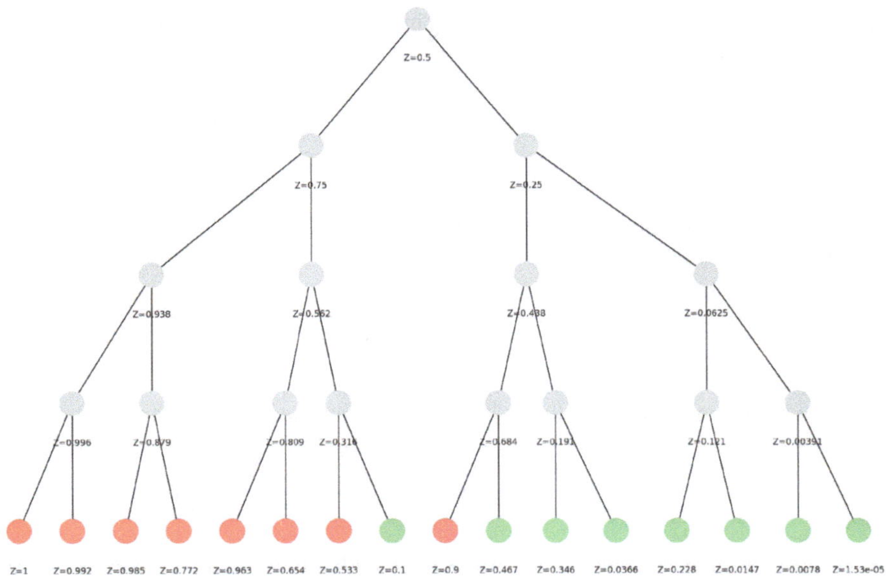

Based on the table derived, the most reliable eight channels (those with the smallest Bhattacharyya parameters) are

$$\boxed{10,\ 11,\ 13,\ 8,\ 12,\ 14,\ 15,\ 16}$$

These are assigned **message bits**. The remaining eight channels are assigned **frozen bits**. The corresponding 5G reliability sequence for $N = 16$ is

$$\boxed{1,\ 2,\ 3,\ 5,\ 9,\ 4,\ 6,\ 10,\ 7,\ 11,\ 13,\ 8,\ 12,\ 14,\ 15,\ 16}$$

From this, the last eight indices (most reliable) are

$$\boxed{7,\ 11,\ 13,\ 8,\ 12,\ 14,\ 15,\ 16}$$

While both methods agree on seven of the most reliable channels, the Bhattacharyya-based approach selects channel 10, whereas the 5G sequence selects channel 7. The discrepancy arises because the 5G reliability sequence is a *universal* empirically optimized order derived through simulation and density evolution, intended to perform well over a wide range of channels and code rates.

[11]

- Block length $N = 32$
- Rate $R = \frac{1}{4} \Rightarrow K = R \cdot N = 8$ information bits

According to the 5G universal reliability sequence, the channels are ordered from least to most reliable. The first 32 elements in the 5G reliability sequence (restricted to values ≤ 32) are:

$$1,\ 2,\ 3,\ 5,\ 9,\ 17,\ 4,\ 6,\ 10,\ 7,\ 18,\ 11,\ 19,\ 13,\ 21,\ 25,\ 8,\ 12,\ 20,\ 14,$$
$$15,\ 22,\ 27,\ 26,\ 23,\ 29,\ 16,\ 24,\ 28,\ 30,\ 31,\ 32$$

The eight most reliable channels are the last eight in this list. Hence, the information bit indices are

$$\boxed{23,\ 29,\ 16,\ 24,\ 28,\ 30,\ 31,\ 32}$$

[12]

For a binary random variable $x \in \{0, 1\}$ for the *reversed* log-likelihood ratio

$$\lambda(x) = \log\left(\frac{P(x=1)}{P(x=0)}\right) = \log\left(\frac{P(x=1)}{1 - P(x=1)}\right)$$

A large *positive* $\lambda(x)$ indicates $x{=}1$ is likely, a large *negative* value favors $x{=}0$, and $\lambda(x) = 0$ reflects maximum uncertainty.

Since $P(x=0) + P(x=1) = 1$,

$$P(x=0) = \frac{1}{1 + e^{\lambda(x)}}, \qquad P(x=1) = \frac{e^{\lambda(x)}}{1 + e^{\lambda(x)}}$$

Let

$$x_1,\ x_2 \in \{0, 1\}, \qquad x_3 \;=\; x_1 \oplus x_2$$

and assume that x_1 and x_2 are statistically independent.

x_1	x_2	$x_3 = x_1 \oplus x_2$
0	0	0
0	1	1
1	0	1
1	1	0

$$\begin{aligned} P(x_3 = 1) &= P(x_1 = 0)P(x_2 = 1) + P(x_1 = 1)P(x_2 = 0) \\ &= \Big(\frac{1}{1 + e^{\lambda_1}}\Big)\Big(\frac{e^{\lambda_2}}{1 + e^{\lambda_2}}\Big) + \Big(\frac{e^{\lambda_1}}{1 + e^{\lambda_1}}\Big)\Big(\frac{1}{1 + e^{\lambda_2}}\Big) \\ &= \frac{e^{\lambda_1} + e^{\lambda_2}}{(1 + e^{\lambda_1})(1 + e^{\lambda_2})} \end{aligned} \tag{B.1}$$

where $\lambda_i \equiv \lambda(x_i)$. Similarly,

$$P(x_3 = 0) = \frac{1 + e^{\lambda_1 + \lambda_2}}{(1 + e^{\lambda_1})(1 + e^{\lambda_2})}.$$

$$\lambda(x_3) \;=\; \log\Big(\frac{P(x_3=1)}{P(x_3=0)}\Big) \;=\; \log\Big(\frac{e^{\lambda_1} + e^{\lambda_2}}{1 + e^{\lambda_1 + \lambda_2}}\Big) = \log\Big(\frac{N}{D}\Big)$$

where we let

$$N = e^{\lambda_1} + e^{\lambda_2}, \qquad D = 1 + e^{\lambda_1 + \lambda_2}.$$

Using the identity,

$$\tanh\Big(-\frac{\lambda}{2}\Big) = \frac{1 - e^{\lambda}}{1 + e^{\lambda}} \implies e^{\lambda} = \frac{1 - \tanh\big(-\frac{\lambda}{2}\big)}{1 + \tanh\big(-\frac{\lambda}{2}\big)}$$

If we define

$$t_1 = \tanh\Big(-\frac{\lambda_1}{2}\Big), \qquad t_2 = \tanh\Big(-\frac{\lambda_2}{2}\Big),$$

then

$$e^{\lambda_1} = \frac{1 - t_1}{1 + t_1}, \qquad e^{\lambda_2} = \frac{1 - t_2}{1 + t_2}$$

Numerator N

$$\begin{aligned} N &= \frac{1 - t_1}{1 + t_1} + \frac{1 - t_2}{1 + t_2} \\ &= \frac{(1 - t_1)(1 + t_2) + (1 - t_2)(1 + t_1)}{(1 + t_1)(1 + t_2)} \\ &= \frac{2\big(1 - t_1 t_2\big)}{(1 + t_1)(1 + t_2)} \end{aligned}$$

Denominator D

$$\begin{aligned} e^{\lambda_1+\lambda_2} &= \frac{1 - t_1}{1 + t_1}\,\frac{1 - t_2}{1 + t_2} = \frac{(1 - t_1)(1 - t_2)}{(1 + t_1)(1 + t_2)}, \\ D = 1 + e^{\lambda_1+\lambda_2} &= \frac{(1 + t_1)(1 + t_2) + (1 - t_1)(1 - t_2)}{(1 + t_1)(1 + t_2)} \\ &= \frac{2\big(1 + t_1 t_2\big)}{(1 + t_1)(1 + t_2)} \end{aligned}$$

Insert N and D

$$\lambda(x_3) = \log\!\left(\frac{2\,(1 - t_1 t_2)/(1 + t_1)(1 + t_2)}{2\,(1 + t_1 t_2)/(1 + t_1)(1 + t_2)}\right) = \log\!\left(\frac{1 - t_1 t_2}{1 + t_1 t_2}\right)$$

Recall

$$\operatorname{atanh}(u) = \frac{1}{2}\log\!\left(\tfrac{1+u}{1-u}\right) \implies \log\!\left(\tfrac{1-u}{1+u}\right) = -2\ \operatorname{atanh}(u)$$

Thus,

$$\boxed{\lambda(x_3) = -\,2\ \operatorname{atanh}\Big[t_1 t_2\Big] = -\,2\ \operatorname{atanh}\Big[\tanh\!\big(-\tfrac{\lambda_1}{2}\big)\ \tanh\!\big(-\tfrac{\lambda_2}{2}\big)\Big]}$$

which could be rewritten given that $\tanh(-z) = -\tanh(z)$ as

$$\lambda(x_3) = -\,2\ \operatorname{atanh}\Big[\tanh\!\big(\tfrac{\lambda_1}{2}\big)\ \tanh\!\big(\tfrac{\lambda_2}{2}\big)\Big]$$

An important point to note is that if we had used the **standard** definition of an LLR, then we recover the tanh rule

$$\lambda(x) = \log\Big(\frac{P(x=0)}{P(x=1)}\Big) = -\log\Big(\frac{P(x=1)}{P(x=0)}\Big) \Longrightarrow$$

$$\lambda(x_3) = 2\ \text{atanh}\Big[\tanh(\tfrac{\lambda_1}{2})\ \tanh(\tfrac{\lambda_2}{2})\Big]$$

[13]

Bits $c_1, c_2, c_3 \in \{0, 1\}$ are *independent.* XOR is associative, so

$$c_1 \oplus c_2 \oplus c_3 = (c_1 \oplus c_2) \oplus c_3, \qquad \text{and let } y = c_1 \oplus c_2$$

The standard definition of a log-likelihood ratio (LLR) is

$$\lambda(x) = \ln\left(\frac{P(x=0)}{P(x=1)}\right)$$

Let

$$y = c_1 \oplus c_2$$

Then

$$\lambda(y) = 2\tanh^{-1}\left[\tanh\left(\frac{\lambda(c_1)}{2}\right)\tanh\left(\frac{\lambda(c_2)}{2}\right)\right]$$

The LLR for $y \oplus c_3$ is given by

$$\lambda(y \oplus c_3) = 2\tanh^{-1}\left[\tanh\left(\frac{\lambda(y)}{2}\right)\tanh\left(\frac{\lambda(c_3)}{2}\right)\right]$$

Substituting the expression for $\lambda(y)$

$$\lambda(c_1 \oplus c_2 \oplus c_3)$$
$$= 2\tanh^{-1}\left[\tanh\left(\frac{1}{2}\cdot 2\tanh^{-1}\left(\tanh\frac{\lambda(c_1)}{2}\tanh\frac{\lambda(c_2)}{2}\right)\right)\tanh\left(\frac{\lambda(c_3)}{2}\right)\right]$$

$$\boxed{= 2\tanh^{-1}\left[\left(\tanh\frac{\lambda(c_1)}{2}\tanh\frac{\lambda(c_2)}{2}\right)\tanh\frac{\lambda(c_3)}{2}\right]}$$

REVERSED LLR You can use the same method, but an alternative approach is as follows.

$$\lambda_i = \ln\!\left(\tfrac{P(c_i=1)}{P(c_i=0)}\right) \implies \left\{\ P(c_i = 1) = \tfrac{e^{\lambda_i}}{1+e^{\lambda_i}},\ P(c_i = 0) = \tfrac{1}{1+e^{\lambda_i}}\right\}$$

$$\boxed{\lambda(c_a \oplus c_b) = -2\ \mathrm{atanh}\Big(\tanh\big(-\lambda_a/2\big)\ \tanh\big(-\lambda_b/2\big)\Big)} \tag{2}$$

Apply (2) with $a = 1,\ b = 2$

$$\lambda(y) = -2\ \mathrm{atanh}\Big(\tanh\big(-\lambda_1/2\big)\ \tanh\big(-\lambda_2/2\big)\Big)$$

Define

$$A = \tanh\big(-\lambda_1/2\big)\ \tanh\big(-\lambda_2/2\big) \quad \implies \quad \tanh\big(-\lambda(y)/2\big) = A$$

Apply (2) with $a = y,\ b = 3$

$$\begin{aligned}\lambda\big((c_1 \oplus c_2) \oplus c_3\big) &= -2\ \mathrm{atanh}\Big(\tanh\big(-\lambda(y)/2\big)\ \tanh\big(-\lambda_3/2\big)\Big)\\ &= -2\ \mathrm{atanh}\big(A\ \tanh\big(-\lambda_3/2\big)\big)\end{aligned}$$

Insert A

$$\boxed{\lambda(c_1 \oplus c_2 \oplus c_3) = -2\ \mathrm{atanh}\Big(\tanh\big(-\lambda_1/2\big)\ \tanh\big(-\lambda_2/2\big)\ \tanh\big(-\lambda_3/2\big)\Big)}$$

[14]

(a) From solution **[6]**,

$$P(c = 0 \mid r) = \frac{f(r \mid c = 0) \cdot P(c = 0)}{f(r)},$$
$$P(c = 1 \mid r) = \frac{f(r \mid c = 1) \cdot P(c = 1)}{f(r)}$$

Taking the ratio, the marginal $f(r)$ cancels out:

$$\frac{P(c = 0 \mid r)}{P(c = 1 \mid r)} = \frac{f(r \mid c = 0)}{f(r \mid c = 1)}$$

From solution **[7]**, we substitute the Gaussian densities

$$f(r \mid c = 0) = \frac{1}{\sigma\sqrt{2\pi}} \exp\left(-\frac{(r - \sqrt{E_s})^2}{2\sigma^2}\right),$$

$$f(r \mid c = 1) = \frac{1}{\sigma\sqrt{2\pi}} \exp\left(-\frac{(r + \sqrt{E_s})^2}{2\sigma^2}\right)$$

Taking their ratio

$$\frac{f(r \mid c = 0)}{f(r \mid c = 1)} = \exp\left(\frac{(r + \sqrt{E_s})^2 - (r - \sqrt{E_s})^2}{2\sigma^2}\right) = \exp\left(\frac{4r\sqrt{E_s}}{2\sigma^2}\right)$$
$$= \exp\left(\frac{2r\sqrt{E_s}}{\sigma^2}\right)$$

Therefore,

$$\lambda = \ln\left(\frac{P(c = 0 \mid r)}{P(c = 1 \mid r)}\right) = \frac{2r\sqrt{E_s}}{\sigma^2}$$

(b) Hence, the log-likelihood ratio can be written in terms of a constant L_c as

$$\lambda_j = L_c r_j, \quad \text{where} \quad L_c = \frac{2\sqrt{E_s}}{\sigma^2}$$

Since SC decoding makes hard decisions based only on the *sign* of λ_j, and λ_j is directly proportional to r_j, the sign of r_j alone is sufficient to determine the bit decision:

$$r_j > 0 \Rightarrow \hat{c}_j = 0, \qquad r_j < 0 \Rightarrow \hat{c}_j = 1$$

Therefore, using $L_{1,j} = r_j$ as the initial belief for each bit $j \in \{1, \ldots, N\}$ is justified in SC decoding because it preserves the sign of the true LLR and hence the decoding decision. It avoids computing or storing the constant scaling factor L_c, simplifying hardware and software implementations, and the decoder performance remains unaffected for hard-decision SC decoding.

However, in other decoding schemes such as list decoding or belief propagation, the actual *magnitude* of the LLR plays a critical role. In these algorithms, the LLR is used to compute path metrics or soft-decision probabilities. Underestimating or ignoring L_c can distort the relative likelihoods, leading to incorrect path pruning (in list decoding) or inaccurate message updates (in belief propagation). For example **LDPC** codes, to be considered in the next chapter, are decoded using the belief propagation (BP) algorithm. There, the log-likelihood ratio $\lambda(c_n|\mathbf{r})$ is computed as

$$\lambda(c_n|\mathbf{r}) = L_c r_n + \log_e\left(\frac{P(c_n = 1 \mid \{r_i,\ i \neq n\})}{P(c_n = 0 \mid \{r_i,\ i \neq n\})}\right)$$

where

- $L_c = \frac{2\sqrt{E_b R_{\text{code}}}}{\sigma^2}$ is the *channel reliability*.
- The term $L_c r_n$ represents the *intrinsic information* from the channel.
- The second term is the *extrinsic information* computed from all other nodes, updated iteratively.

In this case, the scaling factor L_c must be preserved because the *magnitude* of the intrinsic LLR directly influences the weighting between channel evidence and peer messages in the factor graph. In SC decoding, using r_j directly is sufficient since only the sign is needed, but in soft-decision or iterative decoders like LDPC or polar list decoding, the full-scaled LLR must be used to maintain correct reliability weights and convergence behavior.

[15]

Consider a single node in the polar decoding tree where the two code digits are

$$c_1 = u_a \oplus u_b \qquad c_2 = u_b$$

Let the received values (LLRs) from the channel for c_1 and c_2 be

$$r_1 = \ln\frac{P(c_1 = 0)}{P(c_1 = 1)} \qquad r_2 = \ln\frac{P(c_2 = 0)}{P(c_2 = 1)}$$

Assume that the left child u_a has already been decoded and let

$$i = \hat{u}_a \in \{0, 1\}$$

We now determine the decoder's belief about the right child bit u_b, using the available information r_1, r_2 and the value $i = \hat{u}_a$. Namely, to determine the log-likelihood ratio (LLR) for u_b

$$\lambda_g = \ln\left(\frac{P(u_b = 0 \mid r_1, r_2, i)}{P(u_b = 1 \mid r_1, r_2, i)}\right)$$

Recall Bayes' rule

$$P(X \mid Y) = \frac{P(Y \mid X) \cdot P(X)}{P(Y)}$$

Apply it to the numerator and denominator

$$P(u_b = 0 \mid r_1, r_2, i) = \frac{P(r_1, r_2 \mid u_b = 0, i) \cdot P(u_b = 0 \mid i)}{P(r_1, r_2 \mid i)}$$

$$P(u_b = 1 \mid r_1, r_2, i) = \frac{P(r_1, r_2 \mid u_b = 1, i) \cdot P(u_b = 1 \mid i)}{P(r_1, r_2 \mid i)}$$

Take their ratio

$$\frac{P(u_b = 0 \mid r_1, r_2, i)}{P(u_b = 1 \mid r_1, r_2, i)} = \frac{P(r_1, r_2 \mid u_b = 0, i) \cdot P(u_b = 0 \mid i)}{P(r_1, r_2 \mid u_b = 1, i) \cdot P(u_b = 1 \mid i)}$$

Assuming equal priors,

$$P(u_b = 0 \mid i) = P(u_b = 1 \mid i) = \frac{1}{2}$$

Thus,

$$\frac{P(u_b = 0 \mid r_1, r_2, i)}{P(u_b = 1 \mid r_1, r_2, i)} = \frac{P(r_1, r_2 \mid u_b = 0, i)}{P(r_1, r_2 \mid u_b = 1, i)}$$

Taking the logarithm gives

$$\lambda_g = \ln\left(\frac{P(r_1, r_2 \mid u_b = 0, i)}{P(r_1, r_2 \mid u_b = 1, i)}\right)$$

Note that both c_1 and c_2 are completely determined once u_b and u_a are known

$$c_1 = u_a \oplus u_b = i \oplus u_b, \qquad c_2 = u_b$$

Therefore, conditioning on u_b and $u_a = i$ is equivalent to conditioning on specific values of c_1 and c_2, and therefore

$$P(r_1, r_2 \mid u_b, i) = P(r_1, r_2 \mid c_1 = i \oplus u_b,\ c_2 = u_b)$$

Since r_1 and r_2 are the channel observations corresponding to the code bits c_1 and c_2 respectively, and the channel is memoryless (e.g., AWGN), these observations are statistically independent given c_1 and c_2.

$$P(r_1, r_2 \mid c_1, c_2) = P(r_1 \mid c_1) \cdot P(r_2 \mid c_2)$$

Substituting back,

$$P(r_1, r_2 \mid u_b, i) = P(r_2 \mid u_b) \cdot P(r_1 \mid c_1 = i \oplus u_b)$$

So the likelihood ratio becomes

$$\lambda_g = \ln\left(\frac{P(r_2 \mid u_b = 0) \cdot P(r_1 \mid c_1 = i)}{P(r_2 \mid u_b = 1) \cdot P(r_1 \mid c_1 = \bar{i})}\right)$$

From the definition of the standard LLR for a binary variable $x \in \{0, 1\}$ given an observation r, we have

$$\lambda(x) = \ln\left(\frac{P(x=0 \mid r)}{P(x=1 \mid r)}\right)$$

Using Bayes' theorem and assuming equal priors, this becomes

$$\lambda(x) = \ln\left(\frac{P(r \mid x=0)}{P(r \mid x=1)}\right) \Rightarrow \frac{P(r \mid x=0)}{P(r \mid x=1)} = e^{\lambda(x)}$$

In our case, the LLRs are defined as

$$b = \ln\left(\frac{P(r_2 \mid u_b=0)}{P(r_2 \mid u_b=1)}\right) \Rightarrow \frac{P(r_2 \mid u_b=0)}{P(r_2 \mid u_b=1)} = e^{b}$$

and

$$a = \ln\left(\frac{P(r_1 \mid c_1=0)}{P(r_1 \mid c_1=1)}\right) \Rightarrow \frac{P(r_1 \mid c_1=0)}{P(r_1 \mid c_1=1)} = e^{a}$$

If $i = 0$, then $c_1 = i$ for $u_b = 0$ and $c_1 = \bar{i} = 1$ for $u_b = 1$

$$\lambda_g = \ln(e^{b}) + \ln(e^{a}) = b + a$$

If $i = 1$, then $c_1 = i$ for $u_b = 0$ and $c_1 = \bar{i} = 0$ for $u_b = 1$

$$\lambda_g = \ln(e^{b}) + \ln(e^{-a}) = b - a$$

Therefore,

$$\lambda_g = b + (1 - 2i)\,a$$

or equivalently,

$$\boxed{g(a, b, i) = \begin{cases} b + a, & i = 0 \\ b - a, & i = 1 \end{cases}}$$

This is the belief passed to the right child of the SC decoder.

Appendix C
Solutions to Chapter 3

[1] From the given parity-check matrix

$$\mathbf{H} = \begin{bmatrix} 1\,0\,1\,1\,0\,0 \\ 0\,1\,1\,0\,1\,0 \\ 0\,0\,0\,1\,1\,1 \end{bmatrix}$$

The code has length $n = 6$ (number of columns)and $k = 3$ (number of rows). We require the generator matrix $\mathbf{G}$ in the form $[\mathbf{I}_3 \mid \mathbf{P}]$ such that

$$\mathbf{G} \cdot \mathbf{H}^T = \mathbf{0}$$

$$\underbrace{\begin{bmatrix} 1\,0\,0\; p_{11}\; p_{12}\; p_{13} \\ 0\,1\,0\; p_{21}\; p_{22}\; p_{23} \\ 0\,0\,1\; p_{31}\; p_{32}\; p_{33} \end{bmatrix}}_{\mathbf{G}} \cdot \underbrace{\begin{bmatrix} 1\,0\,0 \\ 0\,1\,0 \\ 1\,1\,0 \\ 1\,0\,1 \\ 0\,1\,1 \\ 0\,0\,1 \end{bmatrix}}_{\mathbf{H}^T} = \underbrace{\begin{bmatrix} 0\,0\,0 \\ 0\,0\,0 \\ 0\,0\,0 \end{bmatrix}}_{\mathbf{0}}$$

Solving the equations row by row

$$1 + p_{11} = 0 \Rightarrow p_{11} = 1$$

$$p_{12} = 0$$

$$p_{11} + p_{12} + p_{13} = 0 \Rightarrow 1 + 0 + p_{13} = 0 \Rightarrow p_{13} = 1$$

$$p_{21} = 0$$

$$1 + p_{22} = 0 \Rightarrow p_{22} = 1$$

J. Sodha, *Fundamentals of Polar, LDPC and Turbo Codes*,
https://doi.org/10.1007/978-3-031-90924-5

$$p_{21}+p_{22}+p_{23}=0 \Rightarrow 0+1+p_{23}=0 \Rightarrow p_{23}=1$$
$$1+p_{31}=0 \Rightarrow p_{31}=1$$
$$1+p_{32}=0 \Rightarrow p_{32}=1$$
$$p_{31}+p_{32}+p_{33}=0 \Rightarrow 1+1+p_{33}=0 \Rightarrow p_{33}=0$$

Therefore, the generator matrix is

$$\mathbf{G}=\begin{bmatrix}1&0&0&1&0&1\\0&1&0&0&1&1\\0&0&1&1&1&0\end{bmatrix}$$

[2]

$$\mathbf{H} = \begin{bmatrix}1&1&1&0&0&0\\0&1&1&1&0&0\\0&0&1&1&1&0\end{bmatrix}, \qquad n=6,\ \operatorname{rank}(\mathbf{H})=3 \Longrightarrow k=n-r=3.$$

Let $\mathbf{c}=(c_1,c_2,c_3,c_4,c_5,c_6)$. Foe an all-zero syndrome, we require

$$\text{(R1) } c_1+c_2+c_3=0$$
$$\text{(R2) } c_2+c_3+c_4=0$$
$$\text{(R3) } c_3+c_4+c_5=0$$

For convenience, let

$$u_1=c_1 \qquad u_2=c_2 \qquad u_3=c_6$$

Thus

$$\begin{aligned}
c_3 &= c_1+c_2 && \text{from (R1)}\\
c_4 &= c_2+c_3 = u_2+(u_1+u_2)=u_1 && \text{from (R2)}\\
c_5 &= c_3+c_4 = (u_1+u_2)+u_1=u_2 && \text{from (R3)}
\end{aligned}$$

$$\boxed{\mathbf{c}=\big(u_1,\ u_2,\ u_1+u_2,\ u_1,\ u_2,\ u_3\big)}$$

Now set $(u_1,u_2,u_3)=(1,0,0),\ (0,1,0),\ (0,0,1)$

$$\boxed{\mathbf{G}=\begin{bmatrix}1&0&1&1&0&0\\0&1&1&0&1&0\\0&0&0&0&0&1\end{bmatrix}}$$

As expected,

$$\underbrace{\begin{bmatrix} 1\,0\,1\,1\,0\,0 \\ 0\,1\,1\,0\,1\,0 \\ 0\,0\,0\,0\,0\,1 \end{bmatrix}}\mathbf{G}\;\underbrace{\begin{bmatrix} 1\,0\,0 \\ 1\,1\,0 \\ 1\,1\,1 \\ 0\,1\,1 \\ 0\,0\,1 \\ 0\,0\,0 \end{bmatrix}}\mathbf{H}^{\mathsf{T}} = \begin{bmatrix} 0\,0\,0 \\ 0\,0\,0 \\ 0\,0\,0 \end{bmatrix}.$$

[3]

$$\mathbf{G} = \begin{bmatrix} 1\,0\,0\,0\,1\,1\,0 \\ 0\,1\,0\,0\,1\,0\,1 \\ 0\,0\,1\,0\,0\,1\,1 \\ 0\,0\,0\,1\,1\,1\,1 \end{bmatrix}, \qquad \mathbf{G} = \left[\mathbf{I}_4 \mid \mathbf{P}\right],\; \mathbf{P} = \begin{bmatrix} 1\;1\;0 \\ 1\;0\;1 \\ 0\;1\;1 \\ 1\;1\;1 \end{bmatrix}.$$

Parity-check matrix

$$\mathbf{H} = \left[\mathbf{P}^{\mathsf{T}} \mid \mathbf{I}_3\right] = \begin{bmatrix} 1\;1\;0\;1\;1\;0\;0 \\ 1\;0\;1\;1\;0\;1\;0 \\ 0\;1\;1\;1\;0\;0\;1 \end{bmatrix}.$$

For example, the syndrome for error e_1 in position 1 is

$$\mathbf{s} = \mathbf{e}_1\,\mathbf{H}^{\mathsf{T}} = \underbrace{\left[1\,0\,0\,0\,0\,0\,0\right]}_{1\times 7}\begin{bmatrix} 1\;1\;0 \\ 1\;0\;1 \\ 0\;1\;1 \\ 1\;1\;1 \\ 1\;0\;0 \\ 0\;1\;0 \\ 0\;0\;1 \end{bmatrix} = \left[\,1\,1\,0\,\right]$$

Hence, a single error in position i is the ith row of $\mathbf{H}^T$

bit error position i	row i of $\mathbf{H}^T$	syndrome
1	1 1 0	110
2	1 0 1	101
3	0 1 1	011
4	1 1 1	111
5	1 0 0	100
6	0 1 0	010
7	0 0 1	001

[4]
Augment H with the 5×5 identity and perform row operations mod2. At each stage, the left block is labeled $\mathbf{H}^{(t)}$ and the right block $\mathbf{R}^{(t)}$.

$$\left[\, \mathbf{H}^{(0)} \mid \mathbf{R}^{(0)} \right] = \left[\begin{array}{c|c} 1\,1\,1\,0\,0\,1\,1\,0\,0\,1 & 1\,0\,0\,0\,0 \\ 1\,0\,1\,0\,1\,1\,0\,1\,1\,0 & 0\,1\,0\,0\,0 \\ 0\,0\,1\,1\,1\,1\,0\,1\,0\,1 & 0\,0\,1\,0\,0 \\ 0\,1\,0\,1\,1\,1\,0\,1\,0\,1 & 0\,0\,0\,1\,0 \\ 1\,1\,0\,1\,0\,0\,1\,1\,0\,0 & 0\,0\,0\,0\,1 \end{array}\right]$$

$$R_2 \leftarrow R_2 + R_1, \; R_5 \leftarrow R_5 + R_1$$

$$\Rightarrow \mathbf{H}^{(1)} = \begin{bmatrix} 1\,1\,1\,0\,0\,1\,1\,0\,0\,1 \\ 0\,1\,0\,0\,1\,0\,1\,1\,1\,1 \\ 0\,0\,1\,1\,1\,1\,0\,1\,0\,1 \\ 0\,1\,0\,1\,1\,1\,0\,1\,0\,1 \\ 0\,0\,1\,1\,0\,1\,0\,1\,0\,1 \end{bmatrix}, \quad \mathbf{R}^{(1)} = \begin{bmatrix} 1\,0\,0\,0\,0 \\ 1\,1\,0\,0\,0 \\ 0\,0\,1\,0\,0 \\ 0\,0\,0\,1\,0 \\ 1\,0\,0\,0\,1 \end{bmatrix}$$

$$R_4 \leftarrow R_4 + R_2, \qquad R_1 \leftarrow R_1 + R_2$$

(the second operation is the missing step needed to match the next matrix)

$$\mathbf{H}^{(2)} = \begin{bmatrix} 1\,0\,1\,0\,1\,1\,0\,1\,1\,0 \\ 0\,1\,0\,0\,1\,0\,1\,1\,1\,1 \\ 0\,0\,1\,1\,1\,1\,0\,1\,0\,1 \\ 0\,0\,0\,1\,0\,1\,1\,0\,1\,0 \\ 0\,0\,1\,1\,0\,1\,0\,1\,0\,1 \end{bmatrix}, \; \mathbf{R}^{(2)} = \begin{bmatrix} 0\,1\,0\,0\,0 \\ 1\,1\,0\,0\,0 \\ 0\,0\,1\,0\,0 \\ 1\,1\,0\,1\,0 \\ 1\,0\,0\,0\,1 \end{bmatrix}$$

$$R_1 \leftarrow R_1 + R_3, \; R_5 \leftarrow R_5 + R_3$$

$$\mathbf{H}^{(3)} = \begin{bmatrix} 1\,0\,0\,1\,0\,0\,0\,0\,1\,1 \\ 0\,1\,0\,0\,1\,0\,1\,1\,1\,1 \\ 0\,0\,1\,1\,1\,1\,0\,1\,0\,1 \\ 0\,0\,0\,1\,0\,1\,1\,0\,1\,0 \\ 0\,0\,0\,0\,1\,0\,0\,0\,0\,0 \end{bmatrix}, \; \mathbf{R}^{(3)} = \begin{bmatrix} 0\,1\,1\,0\,0 \\ 1\,1\,0\,0\,0 \\ 0\,0\,1\,0\,0 \\ 1\,1\,0\,1\,0 \\ 1\,0\,1\,0\,1 \end{bmatrix}$$

$$R_1 \leftarrow R_1 + R_4$$

$$\mathbf{H}^{(4)} = \begin{bmatrix} 1\,0\,0\,0\,0\,1\,1\,0\,0\,1 \\ 0\,1\,0\,0\,1\,0\,1\,1\,1\,1 \\ 0\,0\,1\,0\,1\,0\,1\,1\,1\,1 \\ 0\,0\,0\,1\,0\,1\,1\,0\,1\,0 \\ 0\,0\,0\,0\,1\,0\,0\,0\,0\,0 \end{bmatrix}$$

Now eliminate the remaining 1s above the pivot in column 5 using row 5:

$$R_2 \leftarrow R_2 + R_5, \qquad R_3 \leftarrow R_3 + R_5$$

which yields

$$\mathbf{H}^{(5)} = \left[\mathbf{I}_5 \mid \mathbf{P}^\mathsf{T}\right], \qquad \mathbf{P}^\mathsf{T} = \begin{bmatrix} 1\,1\,0\,0\,1 \\ 0\,1\,1\,1\,1 \\ 0\,1\,1\,1\,1 \\ 1\,1\,0\,1\,0 \\ 0\,0\,0\,0\,0 \end{bmatrix}.$$

Hence,

$$\mathbf{P} = (\mathbf{P}^\mathsf{T})^\mathsf{T} = \begin{bmatrix} 1\,0\,0\,1\,0 \\ 1\,1\,1\,1\,0 \\ 0\,1\,1\,0\,0 \\ 0\,1\,1\,1\,0 \\ 1\,1\,1\,0\,0 \end{bmatrix}.$$

Since $\mathbf{H}^{(5)} = [\mathbf{I}_{n-k} \mid \mathbf{P}^\mathsf{T}]$, the corresponding systematic generator (with information bits in the last 5 positions) is

$$\mathbf{G}_{\text{sys}} = \left[\mathbf{P} \mid \mathbf{I}_5\right] = \begin{bmatrix} 1\,0\,0\,1\,0\,1\,0\,0\,0\,0 \\ 1\,1\,1\,1\,0\,0\,1\,0\,0\,0 \\ 0\,1\,1\,0\,0\,0\,0\,1\,0\,0 \\ 0\,1\,1\,1\,0\,0\,0\,0\,1\,0 \\ 1\,1\,1\,0\,0\,0\,0\,0\,0\,1 \end{bmatrix}.$$

Interchanging the first 5 columns with the last 5 (information symbols on the left) gives

$$\boxed{\mathbf{G} = \left[\mathbf{I}_5 \mid \mathbf{P}\right] = \begin{bmatrix} 1\,0\,0\,0\,0\,1\,0\,0\,1\,0 \\ 0\,1\,0\,0\,0\,1\,1\,1\,1\,0 \\ 0\,0\,1\,0\,0\,0\,1\,1\,0\,0 \\ 0\,0\,0\,1\,0\,0\,1\,1\,1\,0 \\ 0\,0\,0\,0\,1\,1\,1\,1\,0\,0 \end{bmatrix}}$$

[5]

For a single polar node with length $N = 2$, the code digits are

$$c_1 = u_1 \oplus u_2, \qquad c_2 = u_2$$

so that

$$c_1 \oplus c_2 = u_1$$

Using the *reversed*-sign LLR

$$\lambda^{(\mathrm{rev})}(x) = \ln\left[\tfrac{P(x=1)}{P(x=0)}\right]$$

and using the two-input tanh rule

$$\lambda_y^{(\mathrm{rev})} = -\,2\ \mathrm{atanh}\left[\tanh\left(-\tfrac{\lambda_1^{(\mathrm{rev})}}{2}\right)\ \tanh\left(-\tfrac{\lambda_2^{(\mathrm{rev})}}{2}\right)\right]$$

For polar codes, the *standard* LLR is used, thus

$$\lambda^{(\mathrm{pol})}(x) = \ln\left[\tfrac{P(x=0)}{P(x=1)}\right] = -\,\lambda^{(\mathrm{rev})}(x)$$

hence, $\lambda_y^{(\mathrm{pol})} = -\lambda_y^{(\mathrm{rev})}$ and

$$\lambda_y^{(\mathrm{pol})} = 2\ \mathrm{atanh}\left[\tanh\left(\tfrac{\lambda_1^{(\mathrm{pol})}}{2}\right)\ \tanh\left(\tfrac{\lambda_2^{(\mathrm{pol})}}{2}\right)\right]$$

Defining $f(\lambda_1, \lambda_2) = \lambda^{(\mathrm{pol})}(c_1 \oplus c_2)$, the left-child update becomes

$$\boxed{f(\lambda_1, \lambda_2) = 2\ \mathrm{atanh}\left[\tanh\left(\tfrac{\lambda_1}{2}\right)\ \tanh\left(\tfrac{\lambda_2}{2}\right)\right]}$$

[6]

$$\lambda(x_i) \;=\; \ln\left(\tfrac{P(x_i=1)}{P(x_i=0)}\right)$$

Thus,

$$P(x_i = 1) = \frac{e^{\lambda(x_i)}}{1 + e^{\lambda(x_i)}} \qquad P(x_i = 0) = \frac{1}{1 + e^{\lambda(x_i)}}$$

$$\begin{aligned} 1 - 2P(x_i = 1) &= 1 - \frac{2e^{\lambda(x_i)}}{1 + e^{\lambda(x_i)}} \\ &= \frac{1 - e^{\lambda(x_i)}}{1 + e^{\lambda(x_i)}} \\ &= \tanh\left(-\tfrac{\lambda(x_i)}{2}\right) \end{aligned}$$

since $\tanh\left(-\frac{u}{2}\right) = \frac{1-e^u}{1+e^u}$. From Gallager's equation

$$P(y=1) = \frac{1}{2}\left[1 - \prod_{i=1}^{N}\left(1 - 2P(x_i = 1)\right)\right]$$

$$\boxed{P(y=1) = \frac{1}{2}\left[1 - \prod_{i=1}^{N} \tanh\left(-\frac{\lambda(x_i)}{2}\right)\right]}$$

[7]

$$\lambda(x_1) = 2.0, \quad \lambda(x_2) = -1.5, \quad \lambda(x_3) = 0.5.$$

For $y = x_1 \oplus x_2 \oplus x_3$,

$$\lambda_y = -2 \ \text{atanh}\left[\tanh\left(-\frac{\lambda(x_1)}{2}\right) \tanh\left(-\frac{\lambda(x_2)}{2}\right) \tanh\left(-\frac{\lambda(x_3)}{2}\right)\right]$$

$$\lambda_y \approx -0.2384$$

$$P(y=1) = \frac{e^{\lambda_y}}{1+e^{\lambda_y}} = \frac{e^{-0.2384}}{1+e^{-0.2384}} \approx 0.441$$

Thus,

$$\boxed{\lambda_y \approx -0.2384 \quad P(y=1) \approx 0.44 \quad P(y=0) \approx 0.56}$$

and since $\lambda_y < 0$, the most likely modulo-2 sum is

$$\hat{y} = 0$$

[8]
From the *reversed* LLR

$$\lambda(c_n \mid \mathbf{r}) = \ln \frac{P(c_n = 1 \mid \mathbf{r})}{P(c_n = 0 \mid \mathbf{r})}, \quad \mathbf{r} = (r_1, \ldots, r_N)$$

By Bayes' rule,

$$P(c_n = 1 \mid \mathbf{r}) = \frac{P(\mathbf{r} \mid c_n = 1)\, P(c_n = 1)}{P(\mathbf{r})}$$

$$P(c_n = 0 \mid \mathbf{r}) = \frac{P(\mathbf{r} \mid c_n = 0)\, P(c_n = 0)}{P(\mathbf{r})}$$

Forming the ratio and cancelling $P(\mathbf{r})$, then assuming $P(c_n = 1) = P(c_n = 0) = \frac{1}{2}$, gives

$$\lambda(c_n \mid \mathbf{r}) = \ln \frac{P(\mathbf{r} \mid c_n = 1)}{P(\mathbf{r} \mid c_n = 0)}$$

Split $\mathbf{r} = (r_n, r_{-n})$ with $r_{-n} = (r_i : i \neq n)$. By the chain rule $P(A, B \mid C) = P(A \mid C)\, P(B \mid A, C)$, we set $A = r_n$, $B = r_{-n}$, $C = c_n$

$$P(\mathbf{r} \mid c_n) = P(r_n \mid c_n)\ P(r_{-n} \mid r_n, c_n)$$

Memoryless AWGN implies

$$P(\mathbf{r} \mid c_n) = P(r_n \mid c_n)\ P(r_{-n} \mid c_n)$$

Thus,

$$\lambda(c_n \mid \mathbf{r}) = \ln \frac{P(r_n \mid c_n = 1)}{P(r_n \mid c_n = 0)} + \ln \frac{P(r_{-n} \mid c_n = 1)}{P(r_{-n} \mid c_n = 0)}$$

Applying Bayes' rule again to the second term,

$$\frac{P(r_{-n} \mid c_n = 1)}{P(r_{-n} \mid c_n = 0)} = \frac{P(c_n = 1 \mid r_{-n})}{P(c_n = 0 \mid r_{-n})}$$

so

$$\lambda(c_n \mid \mathbf{r}) = \ln \frac{P(r_n \mid c_n = 1)}{P(r_n \mid c_n = 0)} + \ln \frac{P(c_n = 1 \mid r_{-n})}{P(c_n = 0 \mid r_{-n})}$$

Let $a = \sqrt{E_b R_{\text{code}}}$. Then

$$P(r_n \mid c_n = 1) = \frac{1}{\sqrt{2\pi\sigma^2}} e^{-\frac{(r_n - a)^2}{2\sigma^2}} \qquad P(r_n \mid c_n = 0) = \frac{1}{\sqrt{2\pi\sigma^2}} e^{-\frac{(r_n + a)^2}{2\sigma^2}}$$

Thus,

$$\ln \frac{P(r_n \mid 1)}{P(r_n \mid 0)} = [-\tfrac{(r_n - a)^2}{2\sigma^2}] - [-\tfrac{(r_n + a)^2}{2\sigma^2}]$$

$$= \frac{(r_n + a)^2 - (r_n - a)^2}{2\sigma^2} = \frac{4\,a\,r_n}{2\sigma^2} = \frac{2\,a}{\sigma^2}\, r_n = L_c\, r_n$$

where $L_c = 2a/\sigma^2$. Therefore,

$$\lambda(c_n \mid \mathbf{r}) = L_c r_n + \ln \frac{P(c_n = 1 \mid r_{-n})}{P(c_n = 0 \mid r_{-n})}$$

or equivalently,

$$\boxed{\lambda(c_n \mid \mathbf{r}) = L_c r_n + \ln \frac{P(c_n = 1 \mid r_i,\ i \neq n)}{P(c_n = 0 \mid r_i,\ i \neq n)}}$$

[9]

The parity-check matrix H

$$H = \begin{bmatrix} 1\,0\,1\,0\,1\,0\,1\,0 \\ 0\,1\,0\,1\,0\,1\,0\,1 \\ 0\,1\,1\,0\,1\,1\,0\,1 \\ 1\,0\,0\,1\,0\,1\,1\,0 \end{bmatrix}$$

$$H = \begin{bmatrix} A & B & T \\ C & D & E \end{bmatrix}$$

Thus,

$$A = \begin{bmatrix} 1\,0\,1\,0 \\ 0\,1\,0\,1 \end{bmatrix} \quad B = \begin{bmatrix} 1\,0 \\ 0\,1 \end{bmatrix} \quad T = \begin{bmatrix} 1\,0 \\ 0\,1 \end{bmatrix}$$

$$C = \begin{bmatrix} 0\,1\,1\,0 \\ 1\,0\,0\,1 \end{bmatrix} \quad D = \begin{bmatrix} 1\,1 \\ 0\,1 \end{bmatrix} \quad E = \begin{bmatrix} 0\,1 \\ 1\,0 \end{bmatrix}$$

$$\mathbf{m} = [1 \quad 0 \quad 1 \quad 1]$$

$$\mathbf{p}_1 = \mathbf{m}(ET^{-1}A + C)^T$$

Since $T = I$,

$$EA = \begin{bmatrix} 0\,1 \\ 1\,0 \end{bmatrix} \begin{bmatrix} 1\,0\,1\,0 \\ 0\,1\,0\,1 \end{bmatrix} = \begin{bmatrix} 0\,1\,0\,1 \\ 1\,0\,1\,0 \end{bmatrix}$$

Then

$$EA + C = \begin{bmatrix} 0\,0\,1\,1 \\ 0\,0\,1\,1 \end{bmatrix} \quad (\text{mod } 2)$$

$$\Rightarrow \mathbf{p}_1 = [1\,0\,1\,1]\begin{bmatrix}0\,0\\0\,0\\1\,1\\1\,1\end{bmatrix} = [1+1 \quad 1+1] = [2 \quad 2] \mod 2 = [0 \quad 0]$$

$$\mathbf{p}_2 = (\mathbf{m}A^T + \mathbf{p}_1 B^T)(T^{-1})^T$$

$$\mathbf{m}A^T = [1\,0\,1\,1]\begin{bmatrix}1\,0\\0\,1\\1\,0\\0\,1\end{bmatrix} = [1+1 \quad 0+1] = [2 \quad 1] \mod 2 = [0 \quad 1]$$

$$\mathbf{p}_1 B^T = [0\,0]\begin{bmatrix}1\,0\\0\,1\end{bmatrix} = [0\,0]$$

$$\Rightarrow \mathbf{p}_2 = [0 \quad 1]$$

$$\mathbf{c} = [\mathbf{m} \quad \mathbf{p}_1 \quad \mathbf{p}_2] = [1 \quad 0 \quad 1 \quad 1 \quad 0 \quad 0 \quad 0 \quad 1]$$

(b) Verification via $\mathbf{c}H^T$

$$\mathbf{c}H^T = [1\,0\,1\,1\,0\,0\,0\,1]\begin{bmatrix}1\,0\,0\,1\\0\,1\,1\,0\\1\,0\,1\,0\\0\,1\,0\,1\\1\,0\,1\,0\\0\,1\,1\,1\\1\,0\,0\,1\\0\,1\,1\,0\end{bmatrix} = [0 \quad 0 \quad 0 \quad 0]$$

Since $\mathbf{c}H^T = \mathbf{0}$, the codeword is valid.
[10]
Bits $c_n \in \{0, 1\}$ are equally likely:

$$P(c_n = 0) = P(c_n = 1) = \tfrac{1}{2}$$

BPSK mapping

$$c_n = 0 \;\mapsto\; s_n = -\sqrt{E_s} \qquad c_n = 1 \;\mapsto\; s_n = +\sqrt{E_s}$$

The received signal point is

$$r_n = s_n + w_n, \quad w_n \sim \mathcal{N}(0, \sigma^2).$$

$$f(r_n \mid c_n = 1) = \frac{1}{\sigma\sqrt{2\pi}} \exp\left[-\frac{(r_n - \sqrt{E_s})^2}{2\sigma^2}\right]$$

$$f(r_n \mid c_n = 0) = \frac{1}{\sigma\sqrt{2\pi}} \exp\left[-\frac{(r_n + \sqrt{E_s})^2}{2\sigma^2}\right]$$

Apply Bayes' rule

$$\begin{aligned} P(c_n = 1 \mid r_n) &= \frac{f(r_n \mid 1)\, P(c_n = 1)}{f(r_n \mid 1)\, P(c_n = 1) + f(r_n \mid 0)\, P(c_n = 0)} \\ &= \frac{f(r_n \mid 1)}{f(r_n \mid 1) + f(r_n \mid 0)} \end{aligned}$$

since $P(c_n = 1) = P(c_n = 0)$

$$P(c_n = 1 \mid r_n) = \frac{1}{1 + \dfrac{f(r_n \mid 0)}{f(r_n \mid 1)}} = \frac{1}{1 + \exp\left[\ln \frac{f(r_n|0)}{f(r_n|1)}\right]}$$

But

$$\frac{f(r_n \mid 0)}{f(r_n \mid 1)} = \exp\left[-\frac{(r_n+\sqrt{E_s})^2-(r_n-\sqrt{E_s})^2}{2\sigma^2}\right] = \exp\left[-\frac{4\, r_n\sqrt{E_s}}{2\sigma^2}\right] = \exp\left[-\frac{2\, r_n\sqrt{E_s}}{\sigma^2}\right]$$

$$\boxed{P(c_n = 1 \mid r_n) = \frac{1}{1 + \exp\left[-2\, r_n\, \frac{\sqrt{E_s}}{\sigma^2}\right]}}$$

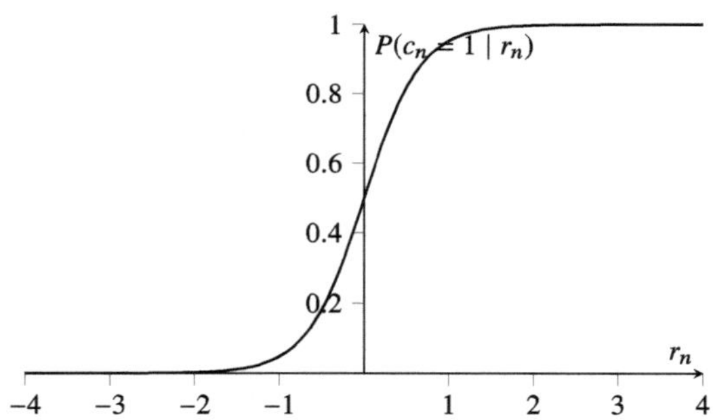

[11]
Given

$$x_3 = x_1 \oplus x_2$$

and

$$\lambda(x_i) = \log_e\left(\frac{P(x_i = 1)}{P(x_i = 0)}\right) \quad \text{for } i = 1, 2$$

x_1	x_2	$x_3 = x_1 \oplus x_2$
0	0	0
0	1	1
1	0	1
1	1	0

From the table

$$P(x_3 = 1) = P(x_1 = 0, x_2 = 1) + P(x_1 = 1, x_2 = 0)$$

Assuming independence between x_1 and x_2

$$P(x_3 = 1) = P(x_1 = 0)P(x_2 = 1) + P(x_1 = 1)P(x_2 = 0)$$

Let

$$p_i = P(x_i = 1) \quad q_i = P(x_i = 0) = 1 - p_i$$

Then

$$\lambda(x_i) = \ln\left(\frac{p_i}{q_i}\right) \quad \Rightarrow \quad e^{\lambda(x_i)} = \frac{p_i}{q_i}$$

Hence,

$$p_i = \frac{e^{\lambda(x_i)}}{1 + e^{\lambda(x_i)}}, \quad q_i = \frac{1}{1 + e^{\lambda(x_i)}}$$

$$\begin{aligned} P(x_3 = 1) &= q_1 p_2 + p_1 q_2 \\ &= \frac{1}{1 + e^{\lambda(x_1)}} \cdot \frac{e^{\lambda(x_2)}}{1 + e^{\lambda(x_2)}} + \frac{e^{\lambda(x_1)}}{1 + e^{\lambda(x_1)}} \cdot \frac{1}{1 + e^{\lambda(x_2)}} \\ &= \boxed{\frac{e^{\lambda(x_2)} + e^{\lambda(x_1)}}{(1 + e^{\lambda(x_1)})(1 + e^{\lambda(x_2)})}} \end{aligned}$$

[12]

$$\lambda_n^{[1]} = L_c r_n + \sum \eta_{m,n}^{[1]}$$

$$L_c = 2$$

$$\lambda_1^{[1]} = L_c r_1 + \eta_{0,1}^{[1]} + \eta_{2,1}^{[1]} + \eta_{3,1}^{[1]} + \eta_{4,1}^{[1]}$$

Using the $\eta_{m,n}$ values:

$$H = \begin{bmatrix} 0\,\mathbf{1}\,1\,1\,0\,1\,0\,1\,0\,1 \\ 1\,0\,1\,0\,1\,0\,1\,0\,1\,0 \\ 0\,\mathbf{1}\,0\,1\,1\,1\,0\,1\,0\,1 \\ 1\,\mathbf{1}\,0\,1\,0\,0\,1\,1\,0\,0 \\ 1\,\mathbf{1}\,0\,1\,0\,0\,1\,1\,0\,0 \end{bmatrix}$$

$$r_1 = -0.605 \quad \Rightarrow \quad L_c r_1 = 2(-0.605) = -1.21$$

$$\lambda_1^{[1]} = -1.21 - 2.1741 - 0.1786 + 0.1920$$

$$= -3.37$$

[13]

Using log-likelihood ratio

$$\lambda(x) = \log_e \left(\frac{P(x=1)}{P(x=0)} \right)$$

- If $\lambda(x) > 0$, then $P(x = 1) > P(x = 0) \Rightarrow$ decode as 1
- If $\lambda(x) < 0$, then $P(x = 0) > P(x = 1) \Rightarrow$ decode as 0

Thus,

$$\lambda_0^{[1]} = -16.403 \quad \Rightarrow 0$$

$$\lambda_1^{[1]} = -3.371 \quad \Rightarrow 0$$

$$\lambda_2^{[1]} = 6.633 \quad \Rightarrow 1$$

$$\lambda_3^{[1]} = 3.611 \quad \Rightarrow 1$$

$$\lambda_4^{[1]} = 7.767 \quad \Rightarrow 1$$

$$\lambda_5^{[1]} = 8.374 \quad \Rightarrow 1$$

$$\lambda_6^{[1]} = -5.403 \quad \Rightarrow 0$$

$$\lambda_7^{[1]} = -4.419 \quad \Rightarrow 0$$

$$\lambda_8^{[1]} = 8.584 \quad \Rightarrow 1$$

$$\lambda_9^{[1]} = -5.14 \quad \Rightarrow 0$$

Decoded codeword $\boxed{0011110010}$

[14]

The parity-check matrix is

$$\mathbf{H} = \begin{bmatrix} 1\,1\,1\,0\,0\,1\,1\,0\,0\,1 \\ 1\,0\,1\,0\,1\,1\,0\,1\,1\,0 \\ 0\,0\,1\,1\,1\,1\,0\,1\,0\,1 \\ 0\,1\,0\,1\,1\,1\,0\,1\,0\,1 \\ 1\,1\,0\,1\,0\,0\,1\,1\,0\,0 \end{bmatrix}$$

and the codeword

$$\mathbf{c} = [c_0, c_1, c_2, c_3, c_4, c_5, c_6, c_7, c_8, c_9]$$

From the second row of **H**, the parity-check equation is

$$c_0 \oplus c_2 \oplus c_4 \oplus c_5 \oplus c_7 \oplus c_8 = 0$$

Solving for c_7

$$c_7 = c_0 \oplus c_2 \oplus c_4 \oplus c_5 \oplus c_8$$

Therefore, the extrinsic information sent from check node 1 to bit node 7 is

$$\eta_{1,7} = \lambda\,(c_7 = 1 \mid r_i, i \neq 7) = \boxed{\lambda\,(c_0 \oplus c_2 \oplus c_4 \oplus c_5 \oplus c_8 = 1 \mid r_i, i \neq 7)}$$

[15]

$$\mathbf{H} = \begin{bmatrix} 1\,1\,1\,0\,0\,1\,\mathbf{1}\,0\,0\,1 \\ 1\,0\,1\,0\,1\,0\,0\,0\,1\,0 \\ 0\,0\,1\,1\,1\,1\,\mathbf{0}\,1\,0\,1 \\ 0\,1\,0\,1\,1\,1\,\mathbf{1}\,1\,0\,1 \\ 1\,1\,0\,1\,0\,0\,\mathbf{1}\,1\,0\,0 \end{bmatrix}$$

$\Rightarrow$ extrinsic terms $\eta_{1,7}, \eta_{3,7}, \eta_{4,7}$

$$\lambda_7^{[1]} = L_c r_7 + \sum \eta_{m,7}^{[1]}$$

Using the parity-check matrix **H**,

$$\lambda_7^{[1]} = L_c r_7 + \eta_{1,7}^{[1]} + \eta_{3,7}^{[1]} + \eta_{4,7}^{[1]}$$

The received value is

$$r_7 = 0.167 \quad \Rightarrow \quad \lambda_7^{[0]} = L_c r_7 = 2(0.167) = 0.334$$

Hence,

$$\lambda_7^{[1]} = L_c r_7 + \eta_{1,7}^{[1]} + \eta_{3,7}^{[1]} + \eta_{4,7}^{[1]}$$

$$= 2(0.167) + (-4.7046 + 0.5992 - 0.6472)$$

$$= \underbrace{0.334}_{\text{intrinsic info.}} + \underbrace{(-4.7046 + 0.5992 - 0.6472)}_{\text{extrinsic info. (dominates to correct)}}$$

$$= -4.4186 \quad \Rightarrow \quad c_7 \text{ now is a zero.}$$

[16]

The decision rule is

$$\lambda_n^{[1]} > 0 \Rightarrow \hat{c}_n = 1, \quad \text{otherwise} \quad \hat{c}_n = 0$$

Thus,

$$\hat{\mathbf{c}} = [0\ 0\ 1\ 1\ 1\ 1\ 0\ 0\ 1\ 0]$$

The syndrome is given by

$$\mathbf{s} = \hat{\mathbf{c}} \cdot \mathbf{H}^{\mathsf{T}}$$

where

$$\mathbf{H}^{\mathsf{T}} = \begin{bmatrix} 1\ 1\ 0\ 0\ 1 \\ 1\ 0\ 0\ 1\ 1 \\ 1\ 1\ 1\ 0\ 0 \\ 0\ 0\ 1\ 1\ 1 \\ 0\ 1\ 1\ 1\ 0 \\ 1\ 0\ 1\ 1\ 0 \\ 1\ 1\ 0\ 0\ 1 \\ 0\ 0\ 1\ 1\ 1 \\ 0\ 1\ 0\ 0\ 0 \\ 1\ 0\ 1\ 1\ 0 \end{bmatrix}$$

$$\mathbf{s} = [0\ 0\ 0\ 1\ 0] \qquad \text{(syndrome after first iteration)}$$

[17]

The parity-check matrix is

$$\mathbf{H} = \begin{bmatrix} 1 & 1 & 0 & 1 & 0 & 0 \\ 0 & 1 & 1 & 0 & 1 & 0 \\ 1 & 0 & 1 & 0 & 0 & 1 \end{bmatrix}$$

The corresponding variable nodes $c_0, c_1, c_2, c_3, c_4, c_5$ and check nodes s_0, s_1, s_2 are shown below in the Tanner graph.

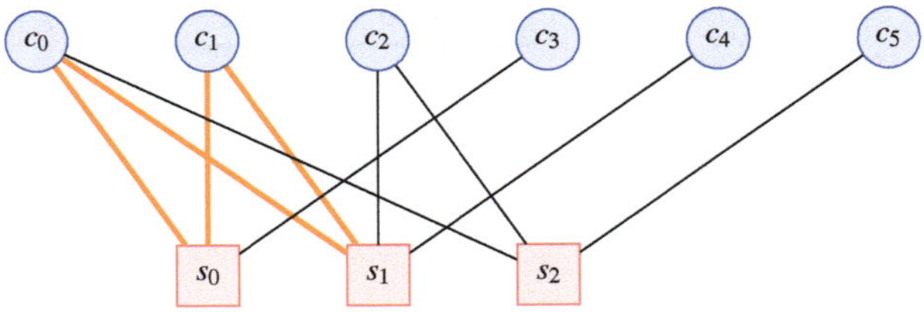

6-cycle highlighted in orange: $c_0 \to s_0 \to c_1 \to s_1 \to c_2 \to s_2 \to c_0$

- A 4-cycle exists if two variable nodes are connected to two common check nodes.

Checking for 4-Cycles:

From the matrix:

$$\text{Row } s_0 : c_0, c_1, c_3$$

$$\text{Row } s_1 : c_1, c_2, c_4$$

$$\text{Row } s_2 : c_0, c_2, c_5$$

Note

- $\mathcal{N}(s_0) \cap \mathcal{N}(s_1) = \{c_1\}$
- $\mathcal{N}(s_0) \cap \mathcal{N}(s_2) = \{c_0\}$
- $\mathcal{N}(s_1) \cap \mathcal{N}(s_2) = \{c_2\}$

Hence

There are no 4-cycles in the Tanner graph corresponding to **H**.

Appendix D
Solutions to Chapter 4

[1] Refer to the trellis diagram shown below. (a) From the traceback path, the decoded information sequence is 1010000. (b) The code word sequence along the traceback path is 11, 10, 00, 10, 11, 00, 00. Compared this with the received sequence 11, 00, 00, 10, 11, 00, 00, we note that the third code digit has mostly likely been corrupted within the channel.

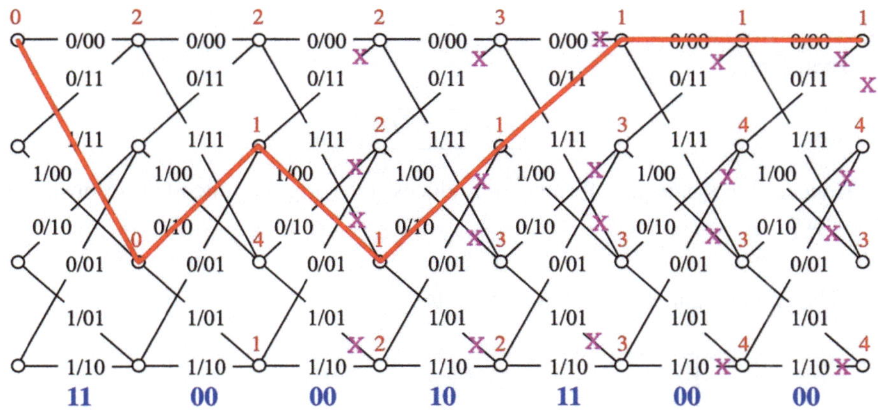

[2]

The rate-$\frac{1}{2}$, constraint-length-3 convolutional encoder generator matrix is

$$G(D) = \left[g_1(D)\ g_2(D)\right] = \left[\ 1 + D + D^2\ ,\ 1 + D^2\right].$$

Let the information sequence be $u(D)$. The encoder produces

$$x^{(1)}(D) = u(D)\, g_1(D) = u(D)\,(1 + D + D^2), \tag{D.1}$$

$$x^{(2)}(D) = u(D)\, g_2(D) = u(D)\,(1 + D^2). \tag{D.2}$$

J. Sodha, *Fundamentals of Polar, LDPC and Turbo Codes*,
https://doi.org/10.1007/978-3-031-90924-5

From the diagram, the tapped connections are

$$h_1(D) = D \quad h_2(D) = 1 + D$$

so the syndrome former output is

$$s(D) = x^{(1)}(D)\,h_1(D) + x^{(2)}(D)\,h_2(D) \pmod 2$$

Thus,

$$\begin{aligned} s(D) &= u(D)\,(1 + D + D^2)\,D \; + \; u(D)\,(1 + D^2)\,(1 + D) \\ &= u(D)\Big[(1 + D + D^2)D + (1 + D^2)(1 + D)\Big] \end{aligned}$$

$$\begin{aligned} (1 + D + D^2)D &= D + D^2 + D^3, \\ (1 + D^2)(1 + D) &= 1 + D + D^2 + D^3, \\ \Rightarrow \; s(D)/u(D) &= (D + D^2 + D^3) + (1 + D + D^2 + D^3) \\ &= 1 \end{aligned}$$

Therefore,

$$\boxed{s(D) = u(D)}$$

Hence, the syndrome former reproduces the original uncoded information sequence.
[3]

(a) The encoder and its state table are as follows.

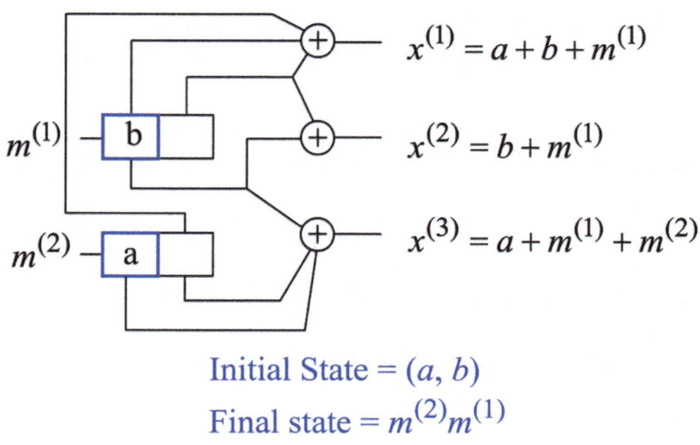

Input $m^{(1)}m^{(2)}$	Initial State (a, b)	Final State $m^{(2)}m^{(1)}$	Code word $x^{(1)}x^{(2)}x^{(3)}$
00	00	00	000
01	00	10	001
10	00	01	111
11	00	11	110
00	01	00	110
01	01	10	111
10	01	01	001
11	01	11	000
00	10	00	101
01	10	10	100
10	10	01	010
11	10	11	011
00	11	00	011
01	11	10	010
10	11	01	100
11	11	11	101

(b) To determine the free distance of this code, we first need to draw its code trellis shown below.

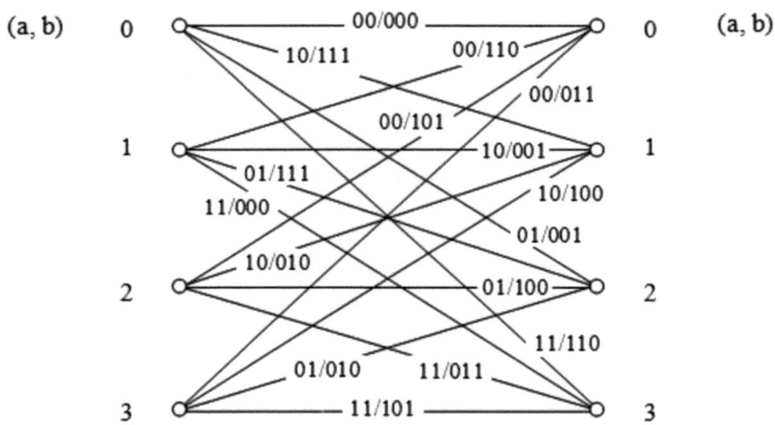

Using the code trellis, we examine all the paths that deviate from the all-zero code word path and merge back with the all-zero path. The path with the minimum code word weight is the free distance of the code. As shown highlighted, the path with the minimum weight is 001, 101, which has a weight of 3. Hence, the free distance of this code is 3.

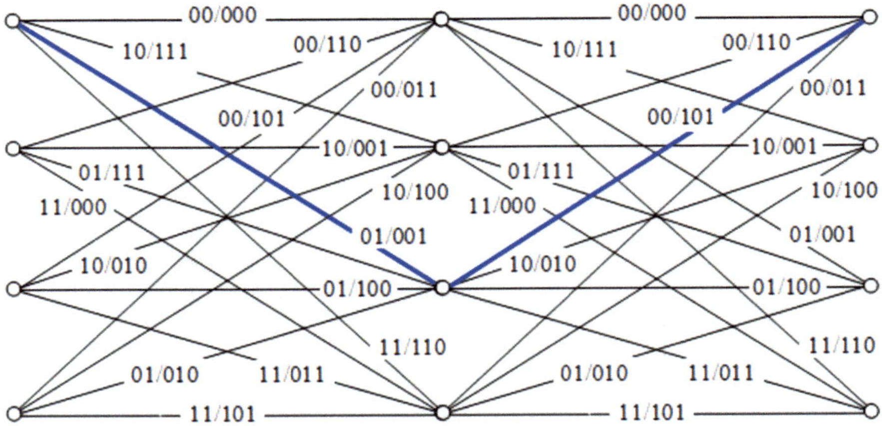

[4]

(a) The encoder and its state table are as follows.

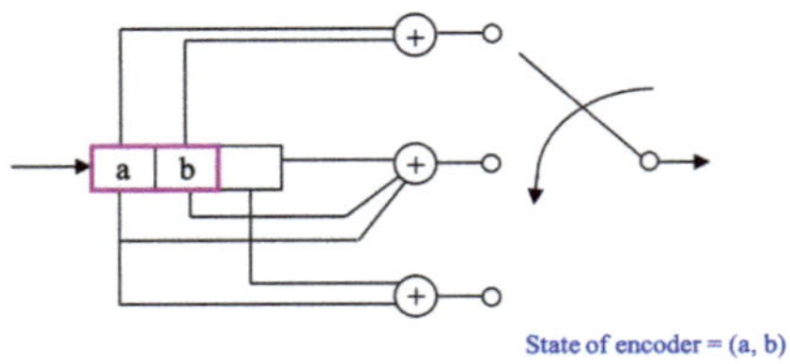

Input	Initial state	Final state	Output
0	00	00	000
1	00	10	111
0	01	00	011
1	01	10	100
0	10	01	110
1	10	11	001
0	11	01	101
1	11	11	010

(b) The free distance is the weight of the highlighted path, which is 7. Examine all other paths that deviate from the all-zero path and merge back, and you will find their weight to be larger.

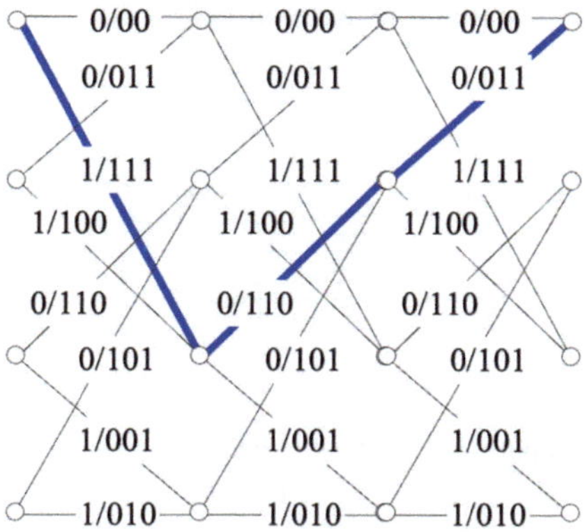

(c) The near-optimum trellis depth would be five times the memory order of this code. Namely, $5(2) = 10$

[5]

(a) $G_1^{(1)}(D) = 1 + D$ and $G_1^{(2)}(D) = 1$ where $G(D) = \left[G_1^{(1)}(D)\ G_1^{(2)}(D) \right] = [1 + D \quad 1]$
The message sequence $m(D) = 1 + D^2 + D^3 + D^4 + \ldots$

$$
\begin{aligned}
x^{(1)}(D) &= m(D)G_1^{(1)}(D) \\
&= (1 + D^2 + D^3 + D^4 + \ldots)(1 + D) \\
&= 1 + D^2 + D^3 + D^4 + \cdots + D + D^3 + D^4 + \ldots \\
&= 1 + D + D^2 \quad \text{or equivalently } 11100000000000\ldots
\end{aligned}
$$

$$
\begin{aligned}
x^{(2)}(D) &= m(D)G_1^{(2)}(D) = (1 + D^2 + D^3 + D^4 + \ldots)(1) \\
&= 1 + D^2 + D^3 + D^4 + \ldots \quad \text{or equivalently } 10111111111\ldots
\end{aligned}
$$

Multiplexing the two sequences shown in the below

11100000000000...
1011111111111 ...

The output of the encoder is 11, 10, 11, 01, 01, 01, 01, ...

(b) The encoder and its state table are as follows.

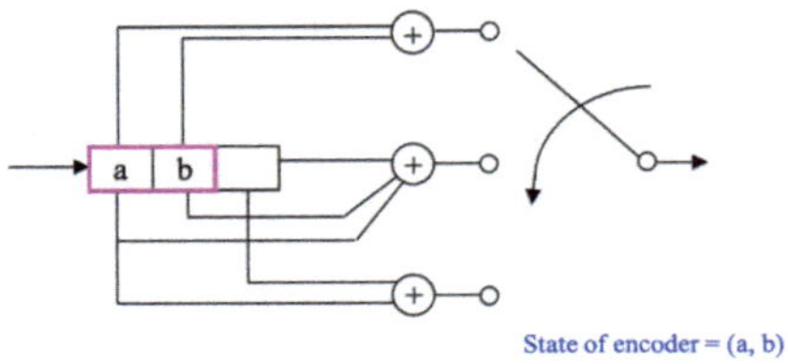

Input	Initial State	Final State	Code Word
0	0	0	00
1	0	1	11
0	1	0	10
1	1	1	01

Codetrellis

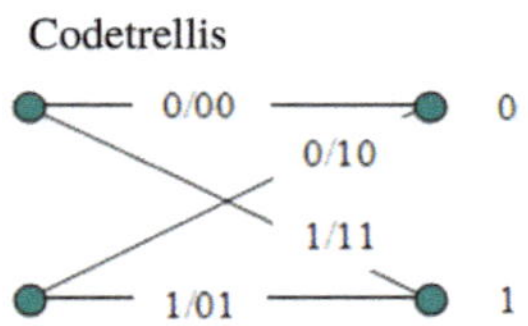

(c)

Encoder Output	11, 10, 11, 01, 01, 01, 01, 01, 01, …
Channel Output	01, 10, 01, 01, 11, 01, 11, 01, 11, …

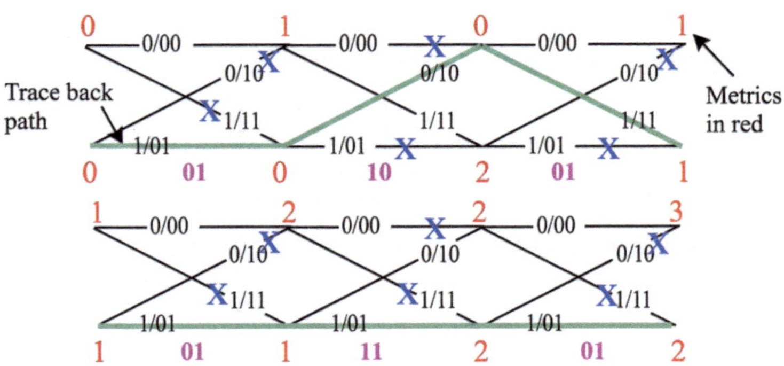

[6]

(a) Input sequence 1011. Starting in the all zero state, output code word sequence is 11,10,00,01,01,11 as shown highlighted in the trellis diagram. Note that two binary digit zeros are inserted to flush the encoder.

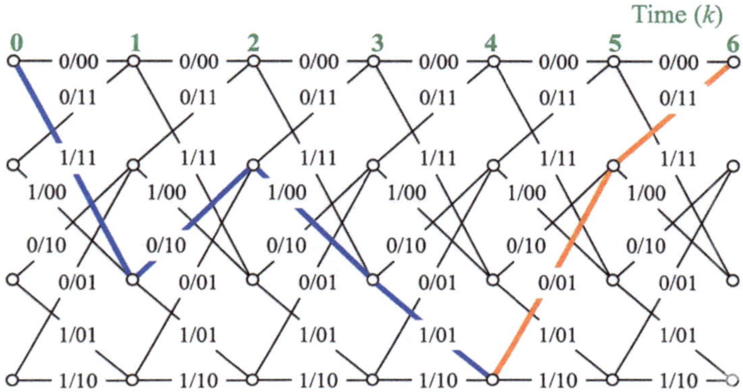

(b) The $\alpha_k(s)$ probabilities are given by

$$\alpha_k(s) = \sum_{s' \in S} \alpha_{k-1}(s')\gamma_k(s', s)$$

where $\gamma_k(s', s) = P(s_k = s, y_k \mid s_{k-1} = s') = \frac{\alpha_{bsc}^h (1-\alpha_{bsc})^{2-h}}{2}$

For $\alpha_{bsc} = 0$ and $h = 0$ we find $\gamma_k(s', s) = \dfrac{1}{2}$
For $\alpha_{bsc} = 0$ and $h \neq 0$ we find $\gamma_k(s', s) = 0$
Thus, the only nonzero probabilities are on the nodes:
$\alpha_0(0), \alpha_1(2), \alpha_2(1), \alpha_3(2), \alpha_4(3), \alpha_5(1), \alpha_6(0)$

$$\alpha_k(0) = \alpha_{k-1}(0)\gamma_k(0, 0) + \alpha_{k-1}(1)\gamma_k(1, 0)$$

$$\alpha_k(1) = \alpha_{k-1}(2)\gamma_k(2, 1) + \alpha_{k-1}(3)\gamma_k(3, 1)$$

$$\alpha_k(2) = \alpha_{k-1}(0)\gamma_k(0, 2) + \alpha_{k-1}(1)\gamma_k(1, 2)$$

$$\alpha_k(3) = \alpha_{k-1}(3)\gamma_k(3, 3) + \alpha_{k-1}(2)\gamma_k(2, 3)$$

$\alpha_0(0) = 1$	0	0	0	0	0		$\alpha_6(0) = \frac{1}{64}$
0	0	0	$\alpha_2(1) = \frac{1}{4}$	0	$\alpha_5(1) = \frac{1}{32}$		0
0	$\alpha_1(2) = \frac{1}{2}$	0	$\alpha_3(2) = \frac{1}{8}$	0	0		0
0	0	0	0	$\alpha_4(3) = \frac{1}{16}$	0		0

$$\beta_{k-1}(s') = \sum_{s \in S} \beta_k(s)\, \gamma_k(s', s)$$

$$\beta_{k-1}(0) = \beta_k(0)\gamma_k(0,0) + \beta_k(2)\gamma_k(0,2)$$

$$\beta_{k-1}(1) = \beta_k(0)\gamma_k(1,0) + \beta_k(2)\gamma_k(1,2)$$

$$\beta_{k-1}(2) = \beta_k(1)\gamma_k(2,1) + \beta_k(3)\gamma_k(2,3)$$

$$\beta_{k-1}(3) = \beta_k(1)\gamma_k(3,1) + \beta_k(3)\gamma_k(3,3)$$

$$\beta_N(s=0) = 1$$

and

$$\beta_N(s \neq 0) = 0$$

$\beta_0(0) = \frac{1}{64}$	0	0	0	0	0	$\beta_6(0) = 1$
0	0	$\beta_2(1) = \frac{1}{16}$	0	0	$\beta_5(1)\frac{1}{2}$	0
0	$\beta_1(2) = \frac{1}{32}$	0	$\beta_3(2) = \frac{1}{8}$	0	0	0
0	0	0	0	$\beta_4(3) = \frac{1}{4}$	0	0

Now

$$\lambda_k(s) = \alpha_k(s)\beta_k(s)$$

$k = 0$	$k = 1$	$k = 2$	$k = 3$	$k = 4$	$k = 5$	$k = 6$
$\lambda_0(0) = \frac{1}{64}$	0	0	0	0	0	$\lambda_6(0) = \frac{1}{64}$
0	0	$\frac{1}{64}$	0	0	$\frac{1}{64}$	0
0	$\frac{1}{64}$	0	$\frac{1}{64}$	0	0	0
0	0	0	0	$\frac{1}{64}$	0	0
Decoded	$u_1 = 1$	$u_2 = 0$	$u_3 = 1$	$u_4 = 1$	$u_5 = 0$	$u_6 = 0$

Using $P\left(u_k = 0 \mid Y_1^N\right) = \frac{1}{\lambda_N(0)} \sum_{S^-} \lambda_k(s)$, we find

$$P\left(u_k = 0 \mid Y_1^6\right) = 64\left[\lambda_k(0) + \lambda_k(1)\right]$$

$$P\left(u_1 = 0 \mid Y_1^6\right) = 64\left[\lambda_1(0) + \lambda_1(1)\right] = 64(0) = 0$$

Since $P\left(u_1 = 0 \mid Y_1^6\right) < 0.5$, $u_1 = 1$.

$$P\left(u_2 = 0 \mid Y_1^6\right) = 64\left[\lambda_2(0) + \lambda_2(1)\right] = 1. \text{ Thus } u_2 = 0.$$

Similarly, $u_3 = 1,\ u_4 = 1,\ u_5 = 0,\ u_6 = 0$.

[7]

Initial state (a, b) and let the final state be (A, B). Then $A = a + b + u_k$ and $B = a$. Also $x_k^{(s)} = u_k$ and $x_k^{(p)} = A + b$.

u_k	Initial state (a, b)	Final state (A, B)	Code word $x_k^{(s)} x_k^{(p)}$
0	00	00	00
1	00	10	11
0	01	10	00
1	01	00	11
0	10	11	01
1	10	01	10
0	11	01	01
1	11	11	10

[8] From the encoder diagram,

$$e = a + d + u_k$$

$$x_k^{(s)} = u_k, \quad x_k^{(p)} = e + b + c + d$$

$$e = a + d + u_k, \quad x_k^{(s)} = u_k, \quad x_k^{(p)} = e + b + c + d$$

u_k	Initial state (a, b, c, d)	Final state (e, a, b, c)	Code word
0	0000	0000	00
1	0000	1000	11
0	0001	1000	00
1	0001	0000	11
0	0010	0001	01
1	0010	1001	10
0	0011	1001	01
1	0011	0001	10
0	0100	0010	01
1	0100	1010	10
0	0101	1010	01
1	0101	0010	10
0	0110	0011	00
1	0110	1011	11
0	0111	1011	00
1	0111	0011	11
0	1000	1100	01
1	1000	0100	10
0	1001	0100	01
1	1001	1100	10
0	1010	1101	00
1	1010	0101	11
0	1011	0101	00
1	1011	1101	11
0	1100	1110	00
1	1100	0110	11
0	1101	0110	00
1	1101	1110	11
0	1110	1111	01
1	1110	0111	10
0	1111	0111	01
1	1111	1111	10

[9]

Final state

$$d = a + b + c + u_k$$

$$x_k^{(s)} = u_k, \quad x_k^{(p)} = d + b + c$$

Input u_k	**Initial State** (a, b, c)	**Final State** (d, a, b)	**Code word** $\left(x_k^{(s)}, x_k^{(p)}\right)$
0	000	000	00
1	000	100	11
0	001	100	00
1	001	000	11
0	010	101	00
1	010	001	11
0	011	001	00
1	011	101	11
0	100	110	01
1	100	010	10
0	101	010	01
1	101	110	10
0	110	011	01
1	110	111	10
0	111	111	01
1	111	011	10

Code trellis

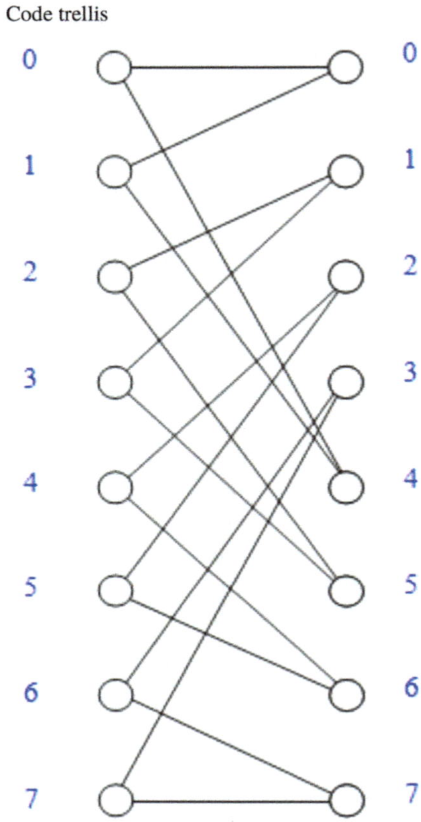

[10]
By definition

$$\gamma_k(s', s) = P(s_k = s, y_k \mid s_{k-1} = s')$$

where $s_{k-1} = s'$ and $s_k = s$ are the previous and current states of the encoder, and y_k is the received symbol at time k. Using the product rule $P(A, B \mid C) = P(A \mid B, C)\, P(B \mid C)$

$$P(s_k = s, y_k \mid s_{k-1} = s') = P(s_k = s \mid s_{k-1} = s') \cdot P(y_k \mid s_k = s, s_{k-1} = s')$$

In a recursive systematic convolutional (RSC) encoder, the transmitted codeword at time k, denoted b_k, is a deterministic function of the transition from state $s_{k-1} = s'$ to state $s_k = s$. Thus

$$P(y_k \mid s_k = s, s_{k-1} = s') = P(y_k \mid b_k)$$

Therefore,

$$\boxed{\gamma_k(s', s) = P(s_k = s \mid s_{k-1} = s') \cdot P(y_k \mid b_k)}$$

[11]

By the chain rule of probability

$$P(s_k = s, Y_1^N) = P(s_k = s, Y_1^k, Y_{k+1}^N)$$

Using the standard product rule $P(A, B) = P(B \mid A)\, P(A)$

$$P(s_k = s, Y_1^k, Y_{k+1}^N) = P(Y_{k+1}^N \mid s_k = s, Y_1^k) \cdot P(s_k = s, Y_1^k)$$

and given the state s_k, the future outputs Y_{k+1}^N are independent of past outputs Y_1^k.

$$P(Y_{k+1}^N \mid s_k = s, Y_1^k) = P(Y_{k+1}^N \mid s_k = s) = \beta_k(s)$$

and

$$P(s_k = s, Y_1^k) = \alpha_k(s)$$

Therefore,

$$\boxed{\lambda_k(s) = \alpha_k(s) \cdot \beta_k(s)}$$

[12]

The normalized forward probability is

$$\widetilde{\alpha}_k(s) = \frac{\sum_{s'} \widetilde{\alpha}_{k-1}(s')\gamma_k(s', s)}{\sum_{s''} \sum_{s'} \widetilde{\alpha}_{k-1}(s')\gamma_k(s', s'')}$$

Let

$$Z_k = \sum_{s''} \sum_{s'} \widetilde{\alpha}_{k-1}(s')\gamma_k(s', s'')$$

Then

$$\widetilde{\alpha}_k(s) = \frac{1}{Z_k} \sum_{s'} \widetilde{\alpha}_{k-1}(s')\gamma_k(s', s)$$

$$\begin{aligned}
\sum_s \widetilde{\alpha}_k(s) &= \sum_s \left(\frac{1}{Z_k} \sum_{s'} \widetilde{\alpha}_{k-1}(s') \gamma_k(s', s) \right) \\
&= \frac{1}{Z_k} \sum_s \sum_{s'} \widetilde{\alpha}_{k-1}(s') \gamma_k(s', s) \\
&= \frac{1}{Z_k} \sum_{s'} \widetilde{\alpha}_{k-1}(s') \sum_s \gamma_k(s', s)
\end{aligned}$$

From the definition

$$Z_k = \sum_{s'} \widetilde{\alpha}_{k-1}(s') \sum_{s''} \gamma_k(s', s'')$$

This is exactly the same expression as in the numerator above. Hence,

$$\boxed{\sum_s \widetilde{\alpha}_k(s) = \frac{Z_k}{Z_k} = 1}$$

[13] Given

$$M_I = 8, \quad P = [17, 37, 19, 29, 41, 23, 13, 7]$$

$$\text{Input Matrix} = \begin{bmatrix} 0 & 1 & 2 & 3 & 4 & 5 & 6 & 7 \\ 8 & 9 & 10 & 11 & 12 & 13 & 14 & 15 \\ 16 & 17 & 18 & 19 & 20 & 21 & 22 & 23 \\ 24 & 25 & 26 & 27 & 28 & 29 & 30 & 31 \\ 32 & 33 & 34 & 35 & 36 & 37 & 38 & 39 \\ 40 & 41 & 42 & 43 & 44 & 45 & 46 & 47 \\ 48 & 49 & 50 & 51 & 52 & 53 & 54 & 55 \\ 56 & 57 & 58 & 59 & 60 & 61 & 62 & 63 \end{bmatrix}$$

For each output position (i, j) in the interleaved matrix

$$\begin{aligned}
x &= \text{mod}(i + j, M_I) \\
P_x &= P[x] \\
i_{\text{row}} &= \text{mod}\left[\left(\frac{M_I}{2} + 1 \right) (i + j), M_I \right] \\
j_{\text{col}} &= \text{mod}\,[(j + 1)P_x - 1, M_I] \\
\text{Interleaved}[i, j] &= \text{Input}[i_{\text{row}}, j_{\text{col}}]
\end{aligned}$$

Compute Interleaved[0][0]

$$
\begin{aligned}
i = 0, \quad j = 0 \Rightarrow x &= \text{mod}(0 + 0, 8) = 0 \\
P_x &= P[0] = 17 \\
i_{\text{row}} &= \text{mod}(5 \cdot 0, 8) = 0 \\
j_{\text{col}} &= \text{mod}(1 \cdot 17 - 1, 8) = \text{mod}(16, 8) = 0 \\
\text{Interleaved}[0][0] &= \text{Input}[0][0] = 0
\end{aligned}
$$

Compute Interleaved[0][1]

$$
\begin{aligned}
i = 0, \quad j = 1 \Rightarrow x &= \text{mod}(0 + 1, 8) = 1 \\
P_x &= P[1] = 37 \\
i_{\text{row}} &= \text{mod}(5 \cdot 1, 8) = \text{mod}(5, 8) = 5 \\
j_{\text{col}} &= \text{mod}(2 \cdot 37 - 1, 8) = \text{mod}(73, 8) = 1 \\
\text{Interleaved}[0][1] &= \text{Input}[5][1] = 41
\end{aligned}
$$

Write			Read		Arrow color
i	j	x	i_{row}	j_{col}	
0	0	0	0	0	
0	1	1	5	1	**Red**
0	2	2	2	0	
0	3	3	7	3	**Orange**
0	4	4	4	4	
0	5	5	1	1	
0	6	6	6	2	
0	7	7	3	7	**Magenta**

Examples

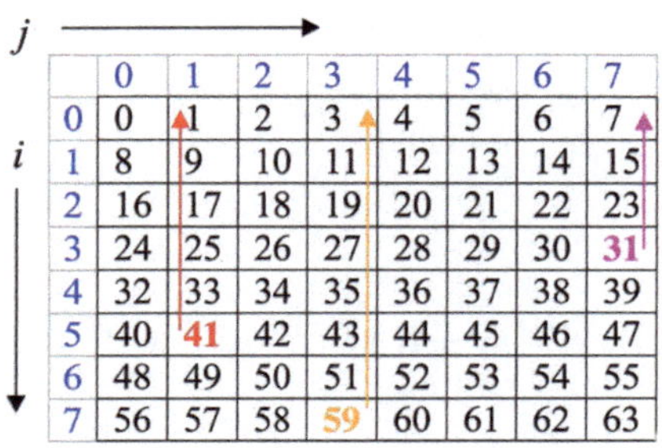

Final Interleaved Matrix is shown below.

$$\text{Interleaved Matrix} = \begin{bmatrix} 0 & 41 & 16 & 59 & 36 & 9 & 50 & 31 \\ 44 & 21 & 62 & 35 & 10 & 53 & 24 & 7 \\ 18 & 57 & 34 & 11 & 48 & 25 & 6 & 47 \\ 60 & 33 & 12 & 51 & 26 & 5 & 42 & 23 \\ 32 & 13 & 54 & 27 & 4 & 45 & 20 & 63 \\ 14 & 49 & 28 & 3 & 40 & 17 & 58 & 39 \\ 52 & 29 & 2 & 43 & 22 & 61 & 38 & 15 \\ 30 & 1 & 46 & 19 & 56 & 37 & 8 & 55 \end{bmatrix}$$

[14]

Gamma values for D1

$$\gamma_k(s', s) = \gamma_k^{(e)}(s', s) \exp\left[\frac{u_k}{2}\left(L^{(e)}(u_k) + L_c y_k^{(s)}\right)\right]$$

$$L_c = 4R\left(\frac{E_b}{N_o}\right), \quad \left(\frac{E_b}{N_o}\right)_{\text{dB}} = 10\log_{10}\left(\frac{E_b}{N_o}\right) = 2\ \text{dB}, \quad \left(\frac{E_b}{N_o}\right) = 10^{2/10}$$

$$\text{Code rate } R = 1/2, \quad L_c = \frac{4}{2}\left(10^{2/10}\right) = 2 \cdot 10^{0.2} = 3.17$$

$$\gamma_k^{(e)}(s', s) = \exp\left[\frac{L_c y_k^{(1p)} x_k^{(1p)}}{2}\right]$$

$$\text{Hence } \gamma_k(s', s) = \exp\left[\frac{L_c y_k^{(1p)} x_k^{(1p)}}{2}\right] \exp\left[\frac{u_k}{2}\left(L^{(e)}(u_k) + L_c y_k^{(1s)}\right)\right]$$

For $\gamma_1(s' = 0, s = 0)$, $y_1^{(1s)} = -1$, $y_1^{(1p)} = -1$,
$x_1^{(1p)} = -1$ (from code trellis diagram),

$$u_1 = -1, \quad L^{(e)}(u_k) = L_{21}^{(e)}(u_D[k]) = 0$$

$$\gamma_1(0, 0) = \exp\left[\frac{L_c(-1)(-1)}{2}\right] \exp\left[\frac{(-1)}{2}(L_c)(-1)\right] = \exp\left[\frac{L_c}{2}\right] \exp\left[\frac{L_c}{2}\right]$$
$$= \exp[L_c]$$

For $\gamma_1(s' = 0, s = 2)$, $y_1^{(1s)} = -1$, $y_1^{(1p)} = -1$, $x_1^{(1p)} = 1$, $u_1 = 1$,

$$L_{21}^{(e)}(u_D[k]) = L^{(e)}(u_k) = 0$$

$$\gamma_1(0,2) = \exp\left[\frac{L_c(-1)(1)}{2}\right]\exp\left[\frac{1}{2}(L_c)(-1)\right] = \exp\left[-\frac{L_c}{2}\right]\exp\left[-\frac{L_c}{2}\right]$$
$$= \exp[-L_c]$$

Clearly $\gamma_1(2,0) = \gamma_1(3,0) = 0$ because the transition from state $s = 2$ or 3 at time $k = 0$ to state $s = 0$ at $k = 1$ is not allowed as evident from the code trellis diagram.

$$\gamma_{15}(s', s) = \exp\left[\frac{L_c y_{15}^{(1p)} x_{15}^{(1p)}}{2}\right]\exp\left[\frac{u_{15}}{2}\left(L_c y_{15}^{(1s)}\right)\right]$$

For $s' = 0, s = 0, y_{15}^{(1s)} = -1, y_{15}^{(1p)} = 0, x_{15}^{(1p)} = -1, u_{15} = -1,$

$$\gamma_{15}(0,0) = \exp\left[\frac{-1}{2}(L_c)(-1)\right] = \exp\left[\frac{L_c}{2}\right]$$

For $s' = 0, s = 2, y_{15}^{(1s)} = -1, y_{15}^{(1p)} = 0, x_{15}^{(1p)} = 1, u_{15} = 1,$

$$\gamma_{15}(0,2) = \exp\left[\frac{1}{2}(L_c)(-1)\right] = \exp\left[-\frac{L_c}{2}\right]$$

For $s' = 1, s = 0, y_{15}^{(1s)} = -1, y_{15}^{(1p)} = 0, x_{15}^{(1p)} = 1, u_{15} = 1,$

$$\gamma_{15}(1,0) = \exp\left[\frac{1}{2}(L_c)(-1)\right] = \exp\left[-\frac{L_c}{2}\right]$$

For $s' = 1$, **and** $s = 2, y_{15}^{(1s)} = -1, y_{15}^{(1p)} = 0, x_{15}^{(1p)} = -1, u_{15} = -1$

$$\gamma_{15}(1,2) = \exp\left[\frac{-1}{2}(L_c(-1))\right] = \exp\left[\frac{L_c}{2}\right]$$

For $s' = 2$, **and** $s = 1, y_{15}^{(1s)} = -1, y_{15}^{(1p)} = 0, x_{15}^{(1p)} = -1, u_{15} = 1$

$$\gamma_{15}(2,1) = \exp\left[\frac{1}{2}(L_c(-1))\right] = \exp\left[-\frac{L_c}{2}\right]$$

[15]

$$\widetilde{\alpha}_k(s) = \frac{\sum_{s'} \widetilde{\alpha}_{k-1}(s')\gamma_k(s', s)}{\sum_s \sum_{s'} \widetilde{\alpha}_{k-1}(s')\gamma_k(s', s)}, \quad \text{thus}$$

$$\widetilde{\alpha}_1(0) = \frac{\sum_{s'} \widetilde{\alpha}_0(s')\gamma_1(s',0)}{\sum_s \sum_{s'} \widetilde{\alpha}_0(s')\gamma_1(s',s)}$$

where

$$\sum_{s'} \widetilde{\alpha}_0(s')\gamma_1(s',0) = \widetilde{\alpha}_0(0)\gamma_1(0,0)+\widetilde{\alpha}_0(1)\gamma_1(1,0)+\widetilde{\alpha}_0(2)\gamma_1(2,0)+\widetilde{\alpha}_0(3)\gamma_1(3,0)$$

But $\widetilde{\alpha}_0(0) = 1$, $\widetilde{\alpha}_0(s \neq 0) = 0$, and from the code trellis, we note that $\gamma_1(2,0) = \gamma_1(3,0) = 0$.

Thus,

$$\sum_{s'} \widetilde{\alpha}_0(s')\gamma_1(s',0) = \gamma_1(0,0)$$

$$\sum_s \sum_{s'} \widetilde{\alpha}_0(s')\gamma_1(s',s) = \sum_s \widetilde{\alpha}_0(0)\gamma_1(0,s) = \gamma_1(0,0)+\gamma_1(0,1)+\gamma_1(0,2)+\gamma_1(0,3)$$

Since $\gamma_1(0,1) = \gamma_1(0,3) = 0$, we find

$$\widetilde{\alpha}_1(0) = \frac{\gamma_1(0,0)}{\gamma_1(0,0) + \gamma_1(0,2)}$$

$$\text{Hence } \widetilde{\alpha}_1(0) = \frac{\exp[L_c]}{\exp[L_c] + \exp[-L_c]} = 0.998$$

$$\widetilde{\alpha}_1(2) = \frac{\sum_{s'} \widetilde{\alpha}_0(s')\gamma_1(s',2)}{\sum_s \sum_{s'} \widetilde{\alpha}_0(s')\gamma_1(s',s)}$$

where

$$\sum_{s'} \widetilde{\alpha}_0(s')\gamma_1(s',2) = \widetilde{\alpha}_0(0)\gamma_1(0,2) = \gamma_1(0,2)$$

and

$$\sum_s \sum_{s'} \widetilde{\alpha}_0(s')\gamma_1(s',s) = \gamma_1(0,0) + \gamma_1(0,2)$$

Note that $\widetilde{\alpha}_1(1) = \widetilde{\alpha}_1(3) = 0$ because the transition from state $s = 0$ at time $k = 0$ to state $s = 1$ or 3 at $k = 1$ is not allowed as evident from the code trellis diagram.

$$\text{Thus } \widetilde{\alpha}_1(2) = \frac{\gamma_1(0,2)}{\gamma_1(0,0)+\gamma_1(0,2)} = \frac{\exp[-L_c]}{\exp[L_c]+\exp[-L_c]} = 1.762 \times 10^{-3}.$$

[16]

$$\widetilde{\beta}_{k-1}(s') = \frac{\sum_s \widetilde{\beta}_k(s)\gamma_k(s',s)}{\sum_s \sum_{s'} \widetilde{\alpha}_{k-1}(s')\gamma_k(s',s)} \quad \text{Thus } \widetilde{\beta}_{15}(s') = \frac{\sum_s \widetilde{\beta}_{16}(s)\gamma_{16}(s',s)}{\sum_s \sum_{s'} \widetilde{\alpha}_{15}(s')\gamma_{16}(s',s)}$$

$$\widetilde{\beta}_{15}(0) = \frac{\sum_s \widetilde{\beta}_{16}(s)\gamma_{16}(0,s)}{\sum_s \sum_{s'} \widetilde{\alpha}_{15}(s')\gamma_{16}(s',s)}$$

$$\sum_s \widetilde{\beta}_{16}(s)\gamma_{16}(0,s) = \widetilde{\beta}_{16}(0)\gamma_{16}(0,0) + \widetilde{\beta}_{16}(1)\gamma_{16}(0,1) + \widetilde{\beta}_{16}(2)\gamma_{16}(0,2) + \widetilde{\beta}_{16}(3)\gamma_{16}(0,3)$$

But $\gamma_k(0,1) = \gamma_k(0,3) = 0$, and with initialization in D1, we have $\beta_N(s \neq 0) = 0$. Thus

$$\sum_s \widetilde{\beta}_{16}(s)\gamma_{16}(0,s) = \gamma_{16}(0,0) = 4.879$$

$$\begin{aligned}
&\sum_s \sum_{s'} \widetilde{\alpha}_{16}(s')\gamma_{16}(s',s) \\
&= \sum_s \widetilde{\alpha}_{15}(0)\gamma_{16}(0,s) + \widetilde{\alpha}_{15}(1)\gamma_{16}(1,s) + \widetilde{\alpha}_{15}(2)\gamma_{16}(2,s) + \widetilde{\alpha}_{15}(3)\gamma_{16}(3,s) \\
&= \widetilde{\alpha}_{15}(0)\gamma_{16}(0,0) + \widetilde{\alpha}_{15}(1)\gamma_{16}(1,0) + \widetilde{\alpha}_{15}(2)\gamma_{16}(2,1) + \widetilde{\alpha}_{15}(3)\gamma_{16}(3,1) \\
&\quad + \widetilde{\alpha}_{15}(0)\gamma_{16}(0,2) + \widetilde{\alpha}_{15}(1)\gamma_{16}(1,2) + \widetilde{\alpha}_{15}(2)\gamma_{16}(2,3) + \widetilde{\alpha}_{15}(3)\gamma_{16}(3,3) \\
&= (0.992 \times 4.879) + (0.002003 \times 0.205) + (0.003918 \times 0.205) + (0.002003 \times 4.879) \\
&\quad + (0.992 \times 0.205) + (0.002003 \times 4.879) + (0.003918 \times 4.879) \\
&\quad + (0.002003 \times 0.205) = 5.084
\end{aligned}$$

Hence,

$$\widetilde{\beta}_{15}(0) = \frac{\sum_{s} \widetilde{\beta}_{16}(s)\gamma_{16}(0, s)}{\sum_{s}\sum_{s'} \widetilde{\alpha}_{15}(s')\gamma_{16}(s', s)} = \frac{4.879}{5.084} = 0.96$$

$$\widetilde{\beta}_{15}(1) = \frac{\sum_{s} \widetilde{\beta}_{16}(s)\gamma_{16}(1, s)}{\sum_{s}\sum_{s'} \widetilde{\alpha}_{16}(s')\gamma_{16}(s', s)} = \frac{\gamma_{16}(1,0)}{5.084} = \frac{0.205}{5.084} = 0.04$$

Finally $\widetilde{\beta}_{15}(2) = \widetilde{\beta}_{15}(3) = 0$ because $\gamma_{16}(2, 0) = \gamma_{16}(3, 0) = 0$.

[17]

To verify the $L_{12}^{(e)}(u_k)$ values for $k = 1, 2$.

$$L_{12}^{(e)}(u_k) = \log_e \left(\frac{\sum_{S^+} \tilde{\alpha}_{k-1}^{(1)}(s')\gamma_k^{(1e)}(s', s)\tilde{\beta}_k^{(1)}(s)}{\sum_{S^-} \tilde{\alpha}_{k-1}^{(1)}(s')\gamma_k^{(1e)}(s', s)\tilde{\beta}_k^{(1)}(s)} \right),$$

where

$$\gamma_k^{(1e)}(s', s) = \exp\left[\frac{L_c y_k^{(1p)} x_k^{(1p)}}{2}\right]$$

and where S^+ and S^- correspond to state transitions by an input information binary digit 1 and 0, respectively.

For $k = 1$,

$$\sum_{S^+} \tilde{\alpha}_0^{(1)}(s')\gamma_1^{(1e)}(s', s)\tilde{\beta}_1^{(1)}(s)$$

$$= \tilde{\alpha}_0^{(1)}(0)\gamma_1^{(1e)}(0, 2)\tilde{\beta}_1^{(1)}(2) + \tilde{\alpha}_0^{(1)}(1)\gamma_1^{(1e)}(1, 0)\tilde{\beta}_1^{(1)}(0) +$$

$$\tilde{\alpha}_0^{(1)}(2)\gamma_1^{(1e)}(2, 1)\tilde{\beta}_1^{(1)}(1) + \tilde{\alpha}_0^{(1)}(3)\gamma_1^{(1e)}(3, 3)\tilde{\beta}_1^{(1)}(3)$$

$$\sum_{S^-} \tilde{\alpha}_0^{(1)}(s')\gamma_1^{(1e)}(s', s)\tilde{\beta}_1^{(1)}(s) = \tilde{\alpha}_0^{(1)}(0)\gamma_1^{(1e)}(0, 0)\tilde{\beta}_1^{(1)}(0) +$$

$$\tilde{\alpha}_0^{(1)}(1)\gamma_1^{(1e)}(1, 2)\tilde{\beta}_1^{(1)}(2) + \tilde{\alpha}_0^{(1)}(2)\gamma_1^{(1e)}(2, 3)\tilde{\beta}_1^{(1)}(3) + \tilde{\alpha}_0^{(1)}(3)\gamma_1^{(1e)}(3, 1)\tilde{\beta}_1^{(1)}(1)$$

But $\tilde{\alpha}_0^{(1)}(0) = 1$ and $\tilde{\alpha}_0^{(1)}(s \neq 0) = 0$.

$$\sum_{S^+} \tilde{\alpha}_0^{(1)}(s')\gamma_1^{(1e)}(s', s)\tilde{\beta}_1^{(1)}(s) = \gamma_1^{(1e)}(0, 2)\tilde{\beta}_1^{(1)}(2)$$

$$\sum_{S^-} \tilde{\alpha}_0^{(1)}(s')\gamma_1^{(1e)}(s',s)\tilde{\beta}_1^{(1)}(s) = \gamma_1^{(1e)}(0,0)\tilde{\beta}_1^{(1)}(0)$$

Thus, for $k = 1$

$$L_{12}^{(e)}(u_1) = \log_e\left(\frac{\sum_{S^+} \tilde{\alpha}_0^{(1)}(s')\gamma_1^{(1e)}(s',s)\tilde{\beta}_1^{(1)}(s)}{\sum_{S^-} \tilde{\alpha}_0^{(1)}(s')\gamma_1^{(1e)}(s',s)\tilde{\beta}_1^{(1)}(s)}\right)$$

$$= \log_e\left(\frac{\tilde{\alpha}_0^{(1)}(0)\gamma_1^{(1e)}(0,2)\tilde{\beta}_1^{(1)}(2)}{\tilde{\alpha}_0^{(1)}(0)\gamma_1^{(1e)}(0,0)\tilde{\beta}_1^{(1)}(0)}\right)$$

$$L_{12}^{(e)}(u_1) = \log_e\left(\frac{\tilde{\alpha}_0^{(1)}(0)\gamma_1^{(1e)}(0,2)\tilde{\beta}_1^{(1)}(2)}{\tilde{\alpha}_0^{(1)}(0)\gamma_1^{(1e)}(0,0)\tilde{\beta}_1^{(1)}(0)}\right)$$

Notice how this log likelihood compares the probability of the transition from state $s' = 0$ at time $k = 0$ to $s = 0$ or $s = 2$ at time $k = 1$.

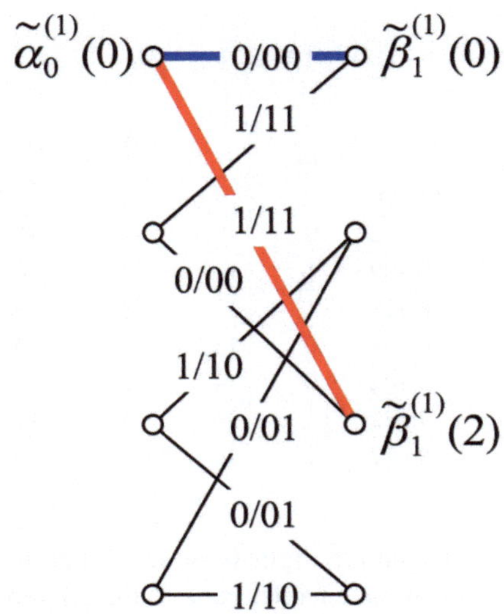

For $k = 1$ and $(s' = 0, s = 0)$, we have $y_1^{(1p)} = -1$, $x_1^{(1p)} = -1$

$$\gamma_1^{(1e)}(0,0) = \exp\left[\frac{L_c y_1^{(1p)} x_1^{(1p)}}{2}\right] = \exp\left[\frac{L_c}{2}\right]$$

For $k = 1$ and $(s' = 0, s = 2)$, we have $y_1^{(1p)} = -1, x_1^{(1p)} = 1$

$$\gamma_1^{(1e)}(0, 2) = \exp\left[\frac{L_c y_1^{(1p)} x_1^{(1p)}}{2}\right] = \exp\left[-\frac{L_c}{2}\right]$$

$$\text{Hence } L_{12}^{(e)}(u_1) = \log_e\left(\frac{\exp\left[-\frac{L_c}{2}\right]\cdot 0.002084}{\exp\left[\frac{L_c}{2}\right]\cdot 0.954}\right) = -9.296$$

To decode the info. sequence, determine

$$L^{(1)}(u_k) = L_c y_k^{(1s)} + L_{21}^{(e)}(u_{D[k]}) + L_{12}^{(e)}(u_k)$$

Since we are not taking into account the contribution from D2, $L_{21}^{(e)}(u_{D[k]}) = 0$, so that

$$L^{(1)}(u_k) = L_c y_k^{(1s)} + L_{12}^{(e)}(u_k)$$

If $L^{(1)}(u_k) > 0$, then $\hat{u}_k = 1$; otherwise, $\hat{u}_k = 0$.

k	$y_k^{(1s)}$	$L_{12}^{(e)}(u_k)$	$L^{(1)}(u_k)$	$\hat{u}_k$	k	$y_k^{(1s)}$	$L_{12}^{(e)}(u_k)$	$L^{(1)}(u_k)$	$\hat{u}_k$
1	−1	−9.296	−12.466	0	9	−1	−6.15	−9.320	0
2	−1	−6.202	−9.372	0	10	−1	−5.513	−8.683	0
3	1	6.239	9.409	1	11	1	6.153	9.323	1
4	1	5.55	8.725	1	12	1	5.533	8.703	1
5	−1	−6.161	−9.331	0	13	−1	−6.164	−9.334	0
6	−1	−5.532	−8.702	0	14	−1	−5.574	−8.744	0
7	1	6.153	9.322	1	15	−1	−6.244	−9.414	0
8	1	5.512	8.682	1	16	−1	−6.205	−9.375	0

Notice how the last two information digits ($k = 15, 16$) correspond to 00, which were necessary to flush the encoder back to the all-zero state.

[18]

For $k = 0$, $\sum_s \tilde{\alpha}_0(s)\tilde{\beta}_0(s) = \tilde{\alpha}_0(0)\tilde{\beta}_0(0) = 0.952$ since $\tilde{\alpha}_0(s \neq 0) = 0$

For $k = 1$,

$$\sum_s \tilde{\alpha}_1(s)\tilde{\beta}_1(s) = \tilde{\alpha}_1(0)\tilde{\beta}_1(0) + \tilde{\alpha}_1(2)\tilde{\beta}_1(2)$$

$$= (0.998 \times 0.954) + (0.001762 \times 0.002084) = 0.952$$

For $k = 16$,

$$\sum_s \tilde{\alpha}_{16}(s)\tilde{\beta}_{16}(s) = \tilde{\alpha}_{16}(0)\tilde{\beta}_{16}(0) = 0.952$$

We conclude that $\sum_s \tilde{\alpha}_k(s)\tilde{\beta}_k(s)$ is independent of k.

References

1. R.E. Ziemer, R.L. Peterson, *Introduction to Digital Communication*, 2nd edn. (Prentice Hall, New York, 2001)
2. C.E. Shannon, A mathematical theory of communication. Bell Syst. Tech. J. **27**(3), 379–423 (1948). https://doi.org/10.1002/j.1538-7305.1948.tb01338.x
3. J.P. Odenwalder, Error control, in *Data Communications, Networks and Systems* (Howard W. Sams, Indianapolis, 1985)
4. J.G. Proakis, *Digital Communications*, 4th edn (McGraw-Hill, New York, 2001), pp. 381–383
5. E. Arikan, Channel polarization: a method for constructing capacity-achieving codes for symmetric binary-input memoryless channels. IEEE Trans. Inf. Theory **55**(7), 3051–3073 (2009). https://doi.org/10.1109/TIT.2009.2021379
6. 3GPP, NR; Multiplexing and channel coding, in *3rd Generation Partnership Project (3GPP), Technical Specification TS 38.212, version 16.2.0, Release 16* (2020)
7. T.K. Moon, *Error Correction Coding: Mathematical Methods and Algorithms*, 2nd edn. (Wiley, New York, 2021). Section 17.3.4. ISBN: 9781119567479
8. R.G. Gallager, Low density parity check codes. IRE Trans. Inform. Theory **8**, 21–28 (1962)
9. T.J. Richardson, A. Shokrollahi, R. Urbanke, Design of capacity-approaching low-density parity-check codes. IEEE Trans. Inform. Theory **47**(2), 619–637 (2001)
10. S. Lin, D.J. Costello, *Error Control Coding*, 2nd edn. (Pearson/Prentice Hall, New York, 2004)
11. A. Viterbi, Error bounds for convolutional codes and an asymptotically optimum decoding algorithm. IEEE Trans. Inform. Theory **13**(2), 260–269 (1967). https://doi.org/10.1109/TIT.1967.1054010
12. J.A. Heller, I.W. Jacobs, Viterbi decoding for satellite and space communication. IEEE Trans. Commun. Technol. **COM-19**(5), 835–848 (1971)
13. L.R. Bahl, J. Cocke, F. Jelinek, J. Raviv, Optimal decoding of linear codes for minimizing symbol error rate. IEEE Trans. Inform. Theory **IT-20**, 248–287 (1974)
14. C. Berrou, A. Glavieux, Near optimum error correcting coding and decoding: turbo-codes. IEEE Trans. Commun. **44**(10), 1261–1271 (1996)
15. W.E. Ryan, A turbo code tutorial, in *New Mexico State University, Box 30001 Dept. 3-O, Las Cruces, NM 88003* (1997). http://www.nmsu.edu
16. E. Hueffmeier, J. Sodha, S. Wicker, Termination Bits in the BCJR Algorithm, in *Proceedings of the Third International Symposium on Communication Systems, Networks and Digital Signal Processing (CSNDSP 2002)* (Staffordshire University, UK, 2002), pp. 101–104

J. Sodha, *Fundamentals of Polar, LDPC and Turbo Codes*,
https://doi.org/10.1007/978-3-031-90924-5

Index

J. Sodha, *Fundamentals of Polar, LDPC and Turbo Codes*,
https://doi.org/10.1007/978-3-031-90924-5

R

S

T

U

V

X

MIX
Papier aus verantwortungsvollen Quellen
Paper from responsible sources
FSC® C105338

If you have any concerns about our products,
you can contact us on
ProductSafety@springernature.com

In case Publisher is established outside the EU,
the EU authorized representative is:
Springer Nature Customer Service Center GmbH
Europaplatz 3, 69115 Heidelberg, Germany

Printed by Libri Plureos GmbH
in Hamburg, Germany